Presented by Schering Health Care

Presented by Schering Health Care

An Atlas of
MULTIPLE SCLEROSIS

THE ENCYCLOPEDIA OF VISUAL MEDICINE SERIES

An Atlas of
MULTIPLE SCLEROSIS

Charles M. Poser, MD, FRCP, FRCP(G)
Department of Neurology, Harvard Medical School
and Beth Israel Deaconess Medical Center
Boston, MA, USA

Foreword by

W. Ian McDonald, PhD, FRCP
Chairman, Department of Clinical Neurological Sciences
National Hospital, Queen Square, and
Harveian Librarian, Royal College of Physicians
London, UK

The Parthenon Publishing Group
International Publishers in Medicine, Science & Technology

NEW YORK   LONDON

Library of Congress Cataloging-in-Publication Data
Poser, Charles M.
      An atlas of multiple sclerosis / Charles M. Poser : foreword by W.
Ian McDonald.
         p. cm. -- (The Encyclopedia of visual medicine series)
      Includes bibliographical references.
      ISBN 1-85070-946-7
      1. Multiple sclerosis--Atlases. I. Title. II. Series.
      [DNLM: 1. Multiple Sclerosis atlases. WL 17 P855a 1998]
RC377.P665 1998
616.8'34'00222--dc21
DNLM/DLC
for Library of Congress   98-17559
                                       CIP

British Library Cataloguing in Publication Data
Poser, Charles M.
    An atlas of multiple sclerosis. - (The encyclopedia of visual medicine
series)
    1. Multiple sclerosis
    I. Title
    616.8'34

    ISBN 1-85070-946-7

Published in the USA by
The Parthenon Publishing Group Inc.
One Blue Hill Plaza
PO Box 1564, Pearl River
New York 10965, USA

Published in the UK and Europe by
The Parthenon Publishing Group Limited
Casterton Hall, Carnforth
Lancs. LA6 2LA, UK

Copyright ©1998
Parthenon Publishing Group

Printed and bound in Spain
by T.G. Hostench, S.A.

To the memory of my mentors

Houston Merritt
and
Ludo van Bogaert

Contents

Foreword

There is a special opportunity for depicting the nervous system and its diseases to combine instructive images with aesthetic satisfaction. Such were the achievements of Carswell and Cruveilhier in the nineteenth century in relation to multiple sclerosis, and of Dawson early in this century. The contributions of these writers were primarily pathological in nature.

Now, Dr Poser has provided us with an atlas that covers the whole range of issues involved in this complex disorder – from its history, through tables illustrating its epidemiology, pictures of its pathology and clinical manifestations to modern methods of diagnosis. We are all in his debt for having gathered together in short compass such a range of fine illustrations which illuminate this succinct and up-to-date account of the disease.

W. Ian McDonald, PhD, FRCP
London

Preface

The past 20 years have seen major changes in the ways multiple sclerosis (MS) is diagnosed and treated. First, computed tomography and, more recently, magnetic resonance imaging (MRI) have introduced clinical imaging techniques into the diagnostic armamentarium of the neurologist. These modalities have also brought new insights to the pathogenesis of the disease. Such mechanodiagnostic procedures, however, have not supplanted the old-fashioned diagnostic methods of clinical history and examination, but rather serve to confirm the diagnosis. Because the neuroimaging changes of MS are non-specific and the problem of MRI differential diagnosis remains difficult, the widely accepted clinical diagnostic criteria published in 1983 remain fundamental to the diagnosis of MS.

Other images are important in the understanding of MS: the topography of the pathological alterations is unique and may be the only means of differentiating the disease from its most common mimic, disseminated encephalomyelitis. Histochemistry, histoimmunology, and electronmicrography have contributed much to our still incomplete understanding of MS pathogenesis. The images produced by these techniques have contributed as much as those from neuroradiology to our diagnostic, investigative, and therapeutic database and, indeed, form an important part of it.

In terms of treatment, although a reduction of exacerbations may result from the use of some of the new drugs, their ability to arrest the progression of the disease remains in question. On the other hand, many advances have been made in the treatment of MS symptoms.

This atlas is dedicated to the memory of two men who were instrumental in guiding my career to the treatment of patients with MS and to research into several of its aspects. Houston Merritt, the ultimate clinician, taught me, as he did so many others, the art of taking a neurological history which, even today, is still the most important component of the MS diagnostic algorithm. The great Belgian neuropathologist Ludo van Bogaert proved to me the truth of Denny-Brown's dictum that "The basic science of neurology is neuropathology." Under his tutelage, with the help of nothing more than the dissecting microscope and exquisite celloidin sections, I learned that MS, while not the only disease of the myelin sheath, leaves its mark in a most distinctive manner.

After more than 40 years of study and treatment of MS, each patient presents a new challenge, each research paper a new idea – or adds more fuel to an existing controversy. Each therapeutic trial brings new hope – and new questions. The riddle of its pathogenesis and the problem of therapy remain unsolved, but the progress that has so far been made in these important areas of research gives much promise for the future.

Acknowledgements

Many colleagues were very kind in contributing images to this atlas; their names appear in the appropriate legends. I wish, however, to express special thanks to the following, who responded rapidly and with extraordinary generosity to my requests:

Dr John Kirk, Queen's University at the Royal Victoria Hospital, Belfast, Northern Ireland;

Professor Ingrid Allen, Queen's University at the Royal Victoria Hospital, Belfast, Northern Ireland;

Dr Frederick Gay, Anglia Polytechnic University, Cambridge, United Kingdom;

Dr Simmons Lessell, Massachusetts Eye and Ear Infirmary, Boston, MA;

Dr Thomas Hedges, Tufts–New England Medical Center, Boston, MA;

Dr William Pendlebury, University of Vermont, Burlington, VT; and

Dr Noble David, University of Miami, Miami, FL.

In addition, Dr Jeffrey Joseph, Beth Israel Deaconess Medical Center, Boston, MA, very kindly reviewed some of the histological images.

Section 1 A Review of Multiple Sclerosis

Introduction

Multiple sclerosis (MS) has fascinated physicians ever since it was first described. Its extraordinary clinical variability and its unpredictability, when added to the complexities of the immune system alterations believed to play an important role in its development, may explain why it has attracted such great interest.

The disease has been known for over one hundred years but, despite enormous expenditures of time and money on research, its pathogenesis is still unresolved. There is, as yet, no clear understanding of the etiological and risk factors, of the meaning of epidemiological data and, most important, of the useful modes of prevention and treatment.

Because it affects men and women in their most productive time of life, and may cause a promising career or a happy marriage to end, it is often viewed as a devastating illness that can rarely be helped by treatment. Whereas the efficacy of specific treatments remains in dispute, there are many symptomatic measures available that can significantly improve the comfort and quality of life of patients with MS. Fortunately, only a minority of patients are rendered severely disabled by the disease.

There is still much to be learned about this disease, but progress has been made, particularly in regard to the reliability of available diagnostic methods as well as in our understanding of its pathology. These are, among others, some of the aspects of MS addressed in this atlas.

Definition

Multiple sclerosis is an inflammatory disease that affects the myelin sheath of the central nervous system (CNS), specifically, the brain and the spinal cord. It is characterized by dissemination in space and in time. In MS, there are lesions involving separate parts of the CNS; signs and symptoms cannot be ascribed to a single lesion. In addition, its clinical course is most often characterized by exacerbations and remissions.

The usual age of onset is in the third and fourth decades. In that age group, MS is second only to epilepsy as the most common disease of the CNS. With rare exceptions, MS does not involve the peripheral nervous system, an important differential feature compared with other diseases primarily involving the myelin sheath.

Historical vignette

The first well-documented instance of MS was the case of Augustus d'Esté, who is often referred to as the nephew of Queen Victoria but was, in fact, the illegitimate grandson of George III. d'Esté's MS started in his second decade of the nineteenth century (Figure 1), and his case was reported in 1948 by Douglas Firth. In 1979, Robert Medaer claimed to have discovered an earlier instance of MS, that of a young Dutchwoman living in the early fourteenth century.

Robert Carswell in London and Jean Cruveilhier in Paris were both, at approximately the same time in the 1830s, the first to publish illustrations of the pathological anatomy of MS (Figures 2–5).

Several clinical descriptions of MS were published in the nineteenth century, but credit is due to Jean Martin Charcot of the Salpêtrière in Paris for the definitive clinical description of the disease in 1868 (Figure 6). Charcot was also the first to correlate the clinical observations with the pathological changes. He believed that what subsequently became known as the 'Charcot triad' – comprising nystagmus, scanning speech and intention tremor – were pathognomonic for the disease.

The next major advance, and the first clue to the pathogenesis of MS, was the demonstration in Edinburgh in 1916 by James Walker Dawson of the intimate relationship between MS plaques and small cerebral blood vessels (Figures 7 and 8). However, it was not until 1947 that Töre Broman of Göteborg, Sweden, discovered what proved to be the crucial step in the development of the MS plaque: the alteration of the blood–brain barrier (BBB) at the edge of the lesion (Figures 9 and 10).

Epidemiology

Multiple sclerosis is not evenly distributed throughout the world. For many years, it had been suggested that the prevalence of MS was directly related to latitude: the further from the equator, the more commonly seen was the disease. Closer examination of this hypothesis, however, showed that this idea was incorrect (Table 1), as there is a wide variation in prevalence among geographical areas of similar latitudes. The major exception to this observation is Tasmania, where the prevalence of MS is double that of southern Australia, despite the fact that the populations in these two areas have almost identical ethnic origins.

Epidemics of MS were said to have occurred in Iceland and in the Faroe Islands as a result of an infection brought by British troops during World War II. Again, closer examination of these data has failed to confirm this idea.

Persons of northern European descent have a strong predisposition for MS. Prevalence and incidence data are often unclear in that no generally accepted diagnostic criteria have been used for inclusion of patients in some studies. In addition, certain surveys have wrongly included possible as well as probable and clinically definite cases.

The north–south gradient that has been repeatedly demonstrated in the United States is derived from the fact that the proportion of the population of Scandinavian descent is considerably higher in the northern tier of states than in the southern. A similar gradient, but at a somewhat lower level of prevalence, is apparent in black American MS patients. This may also be explained by the existence of a gradient, going from the south to the north, that shows an increasing admixture of Caucasian genetic material in black Americans, as measured by blood groups (Figure 11).

Three independent epidemiological studies of MS in immigrants in South Africa, Hawaii, and west coast of the United States have all indicated that the disease is acquired before or at the time of puberty.

Racial or ethnic origin plays an important, but not exclusive, role in determining susceptibility to the disease. No convincingly documented cases of MS have ever been reported in North and South American Indians, Eskimos, Lapps, Australian aborigines, Maoris, Polynesians, Melanesians and Micronesians. The disease is much less frequent in Orientals and extremely rare in black Africans. Because of these indications, numerous studies of genetic markers, in particular, the class II major histocompatibility complex (MHC) alleles of the human leukocyte antigen (HLA) system and their genotypes, have been carried out. These have

Table 1 World distribution of multiple sclerosis: Prevalence and latitude

Location	Latitude	Prevalence*
Iceland	65°N	99.4
Shetland Islands	61°N	129.0
Winnepeg, Canada	50°N	35.0
Seattle, WA, USA	47°N	69.0
Switzerland	47°N	52.0
Parma, Italy	44°N	11.6
Arles, France	44°N	9.0
Krk, Croatia	44°N	44.0
Olmsted County, MN, USA	44°N	122.0
Copparo, Sardinia	44°N	31.0
Asahikawa, Japan	44°N	2.5
Hobart, Tasmania	43°S	68.0
Hautes Pyrenees, France	43°N	39.6
Boston, MA, USA	42°N	41.0
Sassari, Sardinia	41°N	69.0
Alcoy, Spain	39°N	17.0
Seoul, Korea	38°N	2.0
Malta	36°N	4.0
Cape Town, South Africa	36°S	
Afrikaner		10.9
'Colored'/Oriental		3.0
Charleston, SC, USA	33°N	14.0
Newcastle, Australia	33°S	32.5
Israel (native)	32°N	
Sephardi		9.5
Ashkenazi		35.6
New Orleans, LA, USA	30°N	6.0
Kuwait (Arabs)	30°N	
Kuwaiti		9.5
Palestinian		24.0
Canary Islands	29°N	18.3
Okinawa, Japan	26°N	1.9
Hong Kong	23°N	0.8
Bombay (Parsi)	19°N	26.0

*per 100 000 inhabitants

From Poser 1994a, reproduced with permission

shown some fairly strong associations in some populations, such as northern Europeans, but not in others, such as Italians, Israelis and Japanese. The situation in twins is curious and, although the evidence supports the view that a genetic factor is involved, it also raises a few questions: in monozygotic twin pairs, the rate of concordance is approximately only 35%, even when lesions are seen on magnetic resonance imaging (MRI) in the asymptomatic twin.

In addition to the genetic factor, which has been confirmed by numerous studies of twins, siblings, adoptees and family groups, there is also an environmental influence. This has been shown by a number of reliable investigations.

Especially noteworthy is the study in South Africa, where the prevalence of MS is considerably higher in descendants of immigrants of British origin compared with Afrikaners of either French or Dutch descent. The disease is extremely rare in black South Africans.

Frenchmen living in Africa have a considerably lower prevalence rate compared with Frenchmen living in France but, in Martinique in the Caribbean, most of the black patients developing MS have spent some of the critical prepuberal years in metropolitan France. The children of West Indian and Asian immigrants to the United Kingdom are said to have the same level of incidence and prevalence as native-born Englishmen, and the Israelborn children of both Ashkenazic and Sephardic Jews appear to have the same prevalence rates. Methodological considerations have raised questions as to the reliability of these latter studies.

The situation in Hawaii illustrates what appears to be the interplay between genetic and environmental influences. For subjects of Japanese extraction living in either Hawaii or the mainland United States, the environment appears to increase the risk of acquiring MS compared with that of living in Japan. For Caucasians, however, being raised in or moving to Hawaii appears to offer some protection against MS (Table 2).

In summary, epidemiological studies have demonstrated the primary importance of genetic factors modified by as yet unrecognized environmental influences.

Table 2 Multiple sclerosis in Hawaii, California and Japan

Location	Ethnic group	Cases (n)	Prevalence*
Hawaii	Japanese[a]	12	6.5
Hawaii	Caucasians[a]	3	10.5
Hawaii	Caucasians[b]	22	34.4
California	Japanese[c]	17	6.7
California	Caucasians[c]		29.9
Japan	Japanese		2.1

*per 100 000 population
[a]born and raised in Hawaii
[b]adult immigrants to Hawaii
[c]born and raised in California

Etiology

MS has been described as 'a disease of unknown etiology', which implies the existence of a single, specific, causal organism. A number of reports have been published implicating the corona and measles viruses, retroviruses such as the human T-cell lymphotrophic virus (HTLV)-I, herpes simplex virus type 6, the canine distemper virus and the 'MS-associated agent'. None of these has been confirmed, but the idea lingers on despite exhaustive searches by competent investigators using sophisticated techniques. From all the information that is currently available on the disease, it is much more likely that MS is the result in a genetically susceptible subject of the activation of the immune system by different viral agents, thereby initiating a pathogenetic cascade that eventually leads to destruction of the myelin sheath. Many steps in this process remain unknown.

Pathogenesis

The precise pathogenetic mechanism of MS remains controversial, but many investigators believe that alterations of the immune system are responsible. Although there is no doubt that these alterations indeed exist, it is possible that some of these changes are, in fact, the result of the disease process rather than its cause, or that a combination of both processes may be present.

The literature on the pathogenesis of MS is not only voluminous and contradictory, but is also characterized by the frequent use of such words as 'may', 'could', 'would' and 'possible'.

Much of our conception of the pathogenesis of MS is based on data derived from experimental allergic encephalomyelitis, considered by many to be the animal model of MS. Although the entity is surely not an exact model, it shares a number of superficial similarities with MS.

There are two pathogenetic steps that have been fully and generally accepted:

(1) The relationship between plaques and parenchymal blood vessels is close, as was well demonstrated 80 years ago by Dawson and, more recently, by Torben Fog in Copenhagen (Figure 12); and

(2) The increase in the permeability of the blood–brain barrier (BBB) is obligatory.

That this alteration of the BBB is primary rather than the result of destruction of the surrounding myelin became clear from the observation of fluorescein leakage from retinal vessels in optic neuritis; because there is no myelin in the retina, the second possibility had to be eliminated (Figure 13).

Several pathogenetic schemes have been offered, including that proposed by the present author (Figure 14). Based on the surprisingly low concordance rate of MS in identical twins, it is possible to postulate the existence of a 'multiple sclerosis trait' (MST) analogous to those found in sickle cell disease and glucose-6-phosphate dehydrogenase deficiency. The trait is a systemic non-pathological condition or, in other words, a disease waiting to happen.

At present, not all of the components of the MST have been defined nor are they possible to detect in the parents, siblings and children of MS patients. Constituents of the MST include first a vigorous antibody response to a great variety of viral antigens. In addition, there is an inflammatory, primarily lymphocytic, infiltration of the vessels and capillaries of the brain parenchyma. This does not lead to changes in the myelin sheaths, but results in minor alterations of the BBB that cannot be demonstrated by gadolinium-enhanced MRI. This vasculitis may allow B lymphocytes to penetrate into the CNS, where they produce oligoclonal bands. These changes have been well demonstrated in the normal white matter of MS patients, first by Adams and his collaborators (Figure 15) and, more recently, by histoimmunological methods, as used by Gay and his coworkers (Figures 16 and 17), and gadolinium studies (Figure 18).

The acquisition of the MST is considered to be the first step in the development of MS. It is the result of a viral antigenic challenge in genetically susceptible subjects from either an infection or a vaccine. Such a challenge is non-specific; it may be in the form of measles in one subject, an adenovirus in another, and a vaccination against mumps in a third. This initial challenge to the immune system appears to lead primarily to a response of the B lymphocytes that produce antibodies and oligoclonal bands in the cerebrospinal fluid (CSF). The mechanism of the resulting vasculitis is unknown, but it most probably increases the vulnerability of the BBB. At the stage of the MST, the parenchyma of the CNS remains intact. The MS disease may never develop in a subject with the MST. The enhanced antibody response to many viruses and the presence of oligoclonal bands in the CSF have been found in healthy siblings, including the non-concordant twin of a monozygotic pair, of MS patients.

With the development of MS, the impermeability of the BBB is diminished. This alteration of the BBB is one of the few well established and generally accepted steps in the pathogenesis of the disease. Several immunological mechanisms have been proposed that lead to inflammation of the blood vessel wall and, in turn, to alteration of the BBB. There is, however, no agreement as to which mechanism is responsible or whether there is indeed only one.

A key role has been suggested for tumor necrosis factor-α, immune complexes and adhesion molecules. The stimulus for this immunological response is probably a second antigenic challenge from either an infection or vaccination and not necessarily by the same agent that resulted in the MST. However, it is likely to be one that shares part of its molecular structure with the primary agent. Activation of the immune system is through molecular mimicry.

An immunological mechanism is, by far, the most common cause of alteration of the BBB in MS. As a result of this alteration of the BBB, immunoactive lymphocytes penetrate into the brain parenchyma. Many investigators have focused on the as yet poorly understood role of the T lymphocyte in this process. Other immunoactive substances in serum, including complement, interferon-γ and many of the cytokines, as well as B lymphocytes and macrophages, are also able to cross the now permeable BBB and play a still unknown role in the attack on the myelin sheath.

Other agents, which may be termed 'facilitators', such as trauma, electrical injury, organic solvents and vascular accidents, are known to affect the BBB, and may also allow immunoactive lymphocytes and other mediators to penetrate into the parenchyma and affect the myelin. The role of such facilitators remains controversial despite the enormous amount of clinical, pathological, radiological and experimental evidence in its support (Figures 19 and 20). A break in the BBB alone is not sufficient to cause edema or destruction of the myelin sheath. There must be increased activity of the immune system that either induces loss of BBB impermeability or is present when the latter is altered by a facilitator.

Once the immunoactive cells and substances have penetrated into the brain parenchyma, the mechanism by which they injure the myelin sheath is unknown. As no one has ever seen a T cell attacking the myelin sheath, cytokines secreted by these cells have been proposed as the agents of myelinoclasia. The role of the oligodendrocyte in the pathogenetic cascade also remains in dispute. Some neuropathologists have claimed that the disintegration of myelin is secondary to the destruction of oligodendrocytes in early MS lesions, but others have shown that these cells are not affected until later as victims of non-specific destruction.

The primary effects of MS are inflammation and edema, and not destruction of the myelin sheath. In fact, myelin destruction need not necessarily follow (Figure 21). Spontaneous resolution of the inflammation and edema without destruction frequently occurs, and provides a logical explanation for the rapid symptomatic remission seen in some MS patients. Remyelination occurs even in the earliest lesions, but is generally relatively inefficient and much too slow to account for clinical improvement within only a few days.

Myelin is often destroyed and eventually replaced by a glial scar. It is such scars that have given MS its name. The destruction of myelin releases a number of its structural components, including cholesterol, fatty acids, myelin basic protein, myelin-associated glycoprotein, myelin-oligodendrocyte glycoprotein, proteolipid protein, phospholipids, cerebrosides, sphingomyelin and gangliosides. These substances may enter the bloodstream via the permeable BBB and, in turn, provoke an immune response from systemic lymphocytes, thereby creating a vicious circle that results in a self-perpetuating condition. This may also be a possible explanation for the intermittent progression of the disease. Serial MRI studies with gadolinium enhancement have shown that the BBB may remain permeable for an indefinite period of time.

Longitudinal imaging and evoked response studies have shown that the disease is relentlessly progressive even in the absence of clinical exacerbations. This may be better understood if the disease is described as comparable to volcanic island chains, such as Hawaii, where only the tip of the undersea volcanoes are apparent, but with activity continuing unseen beneath the surface of the ocean (Figure 22). It is postulated that, in MS, periods of immune activity, probably stimulated by non-specific viral infections, alternate with periods of immunoquiescence.

Many questions remain unanswered regarding the pathogenetic sequence of MS. The exact role of the various cellular mechanisms leading to the alteration of the BBB and the attack on the myelin sheath has still not been established. The current status of our understanding – or lack thereof – of the role of the immune system in MS has been thus summarized by Cedric Raine as follows:

"In sum, while no single immune system molecule can be assigned as unusual to the CSF of MS, and while there appears to be nothing unique about the manner in which the CNS responds to inflammation, the true uniqueness of the situation in MS is probably related to the many normally sequestered, specific antigens within the myelin sheath and the biology of the myelinating cell, the oligodendrocyte."

Pathology

As MS is a disease of the myelin sheath, lesions are found almost exclusively in the white matter but, because myelinated fibers are present in gray masses, lesions may be seen in gray matter as well. On gross sections of the brain, plaques appear as yellowish, slightly shrunken, areas that are, in fact, dense glial scars (Figures 23–29).

An important characteristic of MS lesions, as seen by various myelin sheath stains, is their very well-defined edge, described as looking as if they have been cut out with a cookie cutter (Figures 30–33). This clearly differentiates the lesions of MS from those of acute disseminated encephalomyelitis. Inflammatory reactions, consisting mostly of lymphocytes, are usually noted around the small blood vessels (Figure 34), especially in acute cases. In the acute phases of the illness, edema is also seen (Figure 35).

Variable amounts of fat-staining lipid material, the so-called myelin abbau, can be observed in macrophages along with myelin fragments (Figures 36 and 37). The intensity of the fat staining is a good index of the age of the lesion: younger lesions contain greater amounts of lipid (Figure 38). Oligodendroglia are not always present in acute lesions, but they will have invariably disappeared from chronic lesions and show deterioration in early lesions.

Large abnormal astrocytes termed 'gemistocytic astrocytes' are often seen (Figure 39) and may form large masses mimicking astrocytic tumors, leading to misinterpretation on biopsy (Figures 40 and 41). Classical MS lesions are periaxonal in that the axon appears unaffected even when the myelin sheath has completely disintegrated. In older, more severe, lesions, however, the axon is atrophic or has disappeared entirely. Necrosis may be seen, especially in acute lesions.

Certain parts of the nervous system are preferentially involved in MS: the anterior angles of the lateral ventricles, the corpus callosum and other periventricular areas are almost invariably affected. Lesions at the periphery of the white matter may extend into the cortex and do not spare the subcortical U-fibers, an important differential from some of the leukodystrophies.

MS lesions vary considerably in size, ranging from only a few millimeters to plaques involving almost the entire centrum semiovale. Asymmetry is the rule and no part of the CNS is spared. Lesions are frequently noted in the spinal cord, especially the cervical portion, as well as, classically, the brain stem and cerebellum, and the cerebral hemispheres (Figures 42–44). The optic nerves and optic chiasm are frequently involved (see Figure 32). The median longitudinal fasciculus is eventually affected in many cases, causing the almost pathognomonic internuclear ophthalmoplegia. The basal ganglia, thalamus and hypothalamus, and dentate nuclei may also be sites of lesions.

In exceptional circumstances, peripheral nerves may become involved. There is often proliferation of Schwann cells into so-called 'onion bulbs', giving the appearance of a Dejerine–Sottas type of neuritis, which may even cause visible and palpable thickening of cutaneous nerves (Figure 45). These nerve lesions are identical to those seen in chronic recurrent Guillain–Barré syndrome and are probably the result of intermittent antigenic challenges which, in the CNS, cause the appearance of new MS plaques. As regards the cranial nerves, lesions of the sensory nuclei of the trigeminal nerve, the intraparenchymatous portion of the facial nerve and connections of the acoustic nerves are not uncommonly encountered. Tic douloureux, or trigeminal neuralgia, in a subject <40 years of age is virtually pathognomonic of MS.

Certain unusual forms of MS have been recognized, mostly on the basis of their pathological appearance. Baló's disease is characterized by the presence of concentric bands of normal-looking myelin alternating with areas of demyelination (Figures 46 and 47). These areas may be relatively large and may occupy a good part of the centrum semiovale. The clinical course of Baló's disease is usually one of rapid progression. However, similar areas of concentric alternating demyelination have been seen in otherwise unremarkable cases of MS.

Another variant of MS is diffuse sclerosis, or Schilder's disease. True Schilder's disease is extremely rare, and is defined as the presence of only two large ($\geq 2 \times 3$ cm), bilateral, slightly asymmetrical areas of demyelination in the cerebral white matter (Figures 48 and 49). Diffuse sclerosis is usually a disease of children. The very large lesions are often associated with the more typical small disseminated lesions of MS in cases that are designated as transitional or diffuse–disseminated sclerosis. In reality, these conditions are simply variants of MS. Thus, 'Schilder's disease' is a term that has been used to describe a heterogeneous group of entities characterized by extensive loss of myelin, including postinfectious and postvaccination encephalomyelitis, leukodystrophies, toxic conditions and lipidoses.

The third variant of MS – Devic's disease or neuromyelitis optica – is the subject of dispute. Lesions affect only the optic nerves and spinal cord, and not always simultaneously. Careful examination of such cases indicates that only approximately one-fourth are cases of MS; the rest are most probably instances of postinfectious encephalomyelitis.

Electronmicroscopy of the classical lesion of MS reveals a number of characteristic changes. One of the earliest is separation of the myelin lamellae by edema. This is often followed by the appearance of macrophages containing myelin debris or fragments, abrupt tapering of the myelin sheath and axonal denudation. Evidence of remyelination may be seen even in the earliest lesions (Figure 50).

Physiology

Normal motor and sensory function is dependent upon the rapid propagation of the nerve impulse along myelinated nerve fibers, which is measured in milliseconds. The myelin sheath is interrupted at regular intervals by the nodes of Ranvier, where the axon is denuded. Because the axon has a high resistance to the electrical impulse and the speed of conduction is too slow, there is an alternative mechanism called 'saltatory conduction'. This is where the electrical impulse jumps from one node of Ranvier to the next while achieving the required conduction velocity. However, if the distance between the available nodes is too great because of destructon of myelin segments, the impulse cannot bridge the gap and saltatory conduction is either impaired or abolished. The electrical impulse must then travel via the slow axonal route (Figure 51).

Another phenomenon that may be observed in some MS patients is the appearance of signs or symptoms due to slowing of nerve conduction because of elevation of body temperature as a result of either ambient heat or fever. The latter is a common cause of pseudoexacerbations. A body temperature increase of as little as 0.1°C may be sufficient to cause such signs and symptoms, which disappear upon cooling.

Clinical aspects

Signs and symptoms

MS most frequently affects the optic nerve and chiasm, brain stem, cerebellum and spinal cord, especially the cervical portion. The presence of spondylosis often contributes to the formation of plaques in that region. This preferential involvement determines the frequency of the signs and symptoms observed in MS patients (Tables 3 and 4). Because it is often difficult, especially on the basis of patient history, to separate symptoms from signs, the clinical features listed in these tables are presented as symptom / sign combinations.

Determining the exact clinical onset of the disease is important for epidemiological investigations.

Table 3 Frequency (%) of multiple sclerosis clinical features in various countries

Clinical feature	USA (n=25)	Canada (n=54)	Denmark (n=60)	Norway (n=31)	England (n=55)	Germany (n=120)
Remission	68.0	76.5	81.7	74.2	72.7	80.0
Pyramidal tract	100.0	83.3*	100.0*	100.0	98.2	90.0
Ocular	92.0	85.2	75.0	80.6	83.6	76.7
Urinary	52.0*	53.7*	71.7	87.1	92.7*	55.8*
Non-equilibratory	76.0	68.5	81.7	80.6	80.0	69.2
Vibration / position	64.0	64.8	51.7	64.5	78.2	76.7
123 Nystagmus	68.0	37.0*	56.7	67.7	72.7*	39.2*
Paresthesiae	68.0	87.0	83.3	54.8	70.9	61.7
Dysarthria	52.0	33.3	40.0	61.3*	52.7	28.3*
Gait ataxia	60.0	46.3	60.0	67.7	45.5	45.0
Mental / cognitive	44.0	15.7*	48.3*	51.6*	41.8	52.5*
Duration of illness						
Mean (yr)	14.0	6.8	8.3	16.5	12.4	13.1
Median (yr)	11.0	4.0	6.0	15.0	12.0	12.0
Range (mo–yr)	2–32	1–43	2–48	3–37	2–36	<12–45

*significance p = 0.01

Table 4 Initial symptoms, clinical course and predominant clinical category in 461 multiple sclerosis patients*

	Frequency		
	Women (n=279) n (%)	Men (n=182) n (%)	Total (n=461) n (%)
Symptom			
Visual loss in one eye	54 (18)	24 (13)	78 (17)
Double vision	27 (10)	35 (19)	62 (13)
Disturbance of balance and gait	38 (14)	45 (25)	83 (18)
Sensory disturbance in limbs	72 (26)	79 (43)	151 (33)
Sensory disturbance in face	10 (4)	6 (3)	16 (3)
Acute myelitic syndrome	20 (7)	6 (3)	26 (6)
Lhermitte's symptom	7 (3)	6 (3)	13 (3)
Pain	5 (2)	3 (2)	8 (2)
Progressive weakness	27 (9)	18 (8)	45 (10)
Clinical course			
Relapsing and remitting	164 (59)	93 (51)	257 (56)
Chronic progressive	67 (24)	60 (33)	127 (28)
Combined	66 (24)	29 (16)	95 (21)
Benign	39 (14)	16 (9)	55 (12)
Predominant clinical category			
Spinal	128 (46)	134 (74)	262 (57)
Cerebellar	23 (8)	35 (19)	58 (13)
Cerebral	11 (4)	7 (4)	18 (4)

*These patients satisfied the Schumacher *et al.* criteria for clinically definite MS and were seen at the MS Clinic of the University of Western Ontario, London, Canada, between 1972 and 1976

From Paty & Poser, 1984, reproduced with permission

Often, non-specific symptoms such as headache, seizure, dizziness or aches and pains are mentioned as the first clinical manifestation of the illness. These onset symptoms of MS may be divided into those which are 'definite' and those which are 'possible'. These symptoms must last for at least 24 hours.

The definite symptoms include unilateral optic / retrobulbar neuritis, true binocular diplopia, acquired monocular color blindness, oscillopsia, transient scanning speech, transverse myelitis, Lhermitte's symptom, gait ataxia, unilateral dysmetria / intention tremor / incoordination, sensory useless hand syndrome, and transient weakness / paresthesiae of an entire limb. The following are also considered to be definite, but only if the patient is less than 40 years of age: tic douloureux; hemifacial spasm; acute unilateral diminution of hearing; transient acute non-positional vertigo; transient painless urinary retention; and transient painless urinary urgency or incontinence in men.

For the following possible symptoms to be considered markers of MS onset, they must be followed by a definite symptom within 2 years: unilateral facial palsy; organic erectile dysfunction; and painful tonic spasms. Transient painless urinary frequency in men, and transient hemiparesis are acceptable only in patients under 40 years of age.

Clinical course

The course of MS has been variously subclassified, but there are essentially only four major groups:

(1) Classic relapsing and remitting type, which represents well over half the cases;

(2) Progressive from onset, which is relatively uncommon;

(3) Secondary progressive, which follows a relapsing–remitting course; and

(4) 'Burned-out' cases, in which the disease appears to have become arrested. Most of these patients have become wheelchair-bound, but their upper extremities remain functional. This last group is rarely mentioned, yet constitutes approximately 20% of cases.

Longitudinal studies of MS with evoked potential and imaging studies have shown that, even in cases that appear to be clinically inactive, there is intermittent activity and progression of the disease. MS can also follow a hyperacute course, leading to death in a matter of a few weeks; this type has been called 'Marburg disease'. Such cases lacking autopsy confirmation or which are based on a misinterpretation of a needle biopsy are often, in fact, cases of acute monophasic disseminated encephalomyelitis rather than MS. In true Marburg disease, the lesions resemble those of MS, having the characteristic sharply defined edges, and showing different ages and stages of development, in contrast to the ill-defined, uniformly aged, areas of demyelination of disseminated encephalomyelitis.

Most exacerbations of MS appear to follow on the heels of a viral infection. Because such infections may be very mild or, in some cases, with no overt clinical manifestations, the triggering event may not be recognized. The roles of trauma, electrical injury, vaccinations and emotional stress in precipitating attacks of MS remain controversial.

The correlation between the number, site and size of MS lesions – as revealed by neuroimaging and autopsy – and clinical manifestations is poor. Many plaques involve the so-called 'silent' areas of the brain. The reason why some lesions do not cause neurological dysfunction is best explained by the theoretical concept of the 'safety factor'. The disease process must impair conduction in a critical, minimum number of fibers in both motor and sensory tracts before clinical dysfunction occurs. The number of available fibers above this minimum number constitutes the safety factor. If involvement is due to inflammation and edema, the disease may be partially or wholly reversible, or the safety factor may be partially or totally abolished by myelinoclasia, leaving the patient with permanent signs or symptoms (Figure 52).

Diagnosis

Diagnostic criteria

The diagnosis of MS is a clinical exercise based on the characteristic dissemination of the lesions in both time and space. This principle applies to the overwhelming majority of cases. The diagnostic criteria that are now virtually universally adopted were first published in 1983 (Table 5).

Explanation of terms

Attack (bout, episode, exacerbation): Occurrence of a symptom or symptoms of neurological dysfunction, with or without objective confirmation, of >24-hours' duration. This may be completely subjective and anamnestic.

Historical information: Description of symptoms by the patient; medical records confirming anamnestic information should be obtained whenever possible.

Clinical evidence: Abnormal signs demonstrable by neurological examination. Such neurological signs are acceptable even if no longer present, provided that they were previously recorded by a competent examiner.

Paraclinical evidence: Demonstration by neuroimaging or evoked potential studies of CNS lesions that

Table 5 Diagnostic criteria for multiple sclerosis

Category	ATT	CLN	PCL	CSF
Clinical definite				
1	2	2		
2	2	1 and	1	
Laboratory definite				
1	2	1 or	1	+
2	1	2		+
3	1	1 and	1	+
Clinically probable				
1	2	1		
2	1	2		
3	1	1 and	1	
Laboratory-supported probable				
1	2			+

ATT, attack; CLN, clinical evidence; PCL, paraclinical evidence; CSF, CSF oligoclonal bands or increased IgG; +, present

From Poser et al., 1983, with permission

may or may not previously have caused signs or symptoms. Such lesions must be at locations different from those recorded by history or examination.

Typical of MS: Signs and symptoms that indicate involvement of areas of the CNS that are most commonly affected by MS. Examples include optic neuritis, internuclear ophthalmoplegia, intention tremor, spastic–ataxic gait and Lhermitte's symptom. Signs and symptoms reflecting gray matter lesions are so uncommon that they should not be taken into account when establishing the diagnosis. Lesions of spinal roots and nerves may not be counted.

Remission: A definite improvement, lasting for at least a month, in signs or symptoms that have been present for 24 hours or more.

Separate lesions: Clinical signs and symptoms that cannot be explained on the basis of a single lesion.

Laboratory support: Applies only to the examination of CSF for an increased level of immunoglobulin G (IgG) and the presence of oligoclonal bands. Evoked responses and neuroimaging studies provide paraclinical evidence and are not included under this rubric.

Optic fundus and visual fields

Because the optic nerve and the visual system are so frequently involved, and the optic nerve head is the only externally visible part of the CNS, examining the optic fundus and plotting the visual fields provides valuable clinical information (Figures 53–55).

Confirmatory laboratory procedures

With few exceptions, the laboratory procedures described below should be used only if the clinical information and findings on neurological examination are insufficient to warrant making the diagnosis of definite MS. Thus, these procedures are intended to confirm a suspicion of MS or, on rare occasions, rule out conditions such as disseminated encephalomyelitis (in particular, the recurrent and multiphasic types), Lyme disease, HTLV-I-associated paraparesis and sarcoidosis.

Examination of cerebrospinal fluid

Lumbar puncture as an adjunct for the diagnosis of MS has become increasingly rare, but remains an essential procedure when other conditions, such as Lyme disease, sarcoidosis, HTLV-1-associated paraparesis, AIDS, and neurosyphilis, must be ruled out. Thus, it remains a useful confirmatory test for MS.

In an acute attack of MS, there is often a slight elevation of the white blood cell count – mostly of lymphocytes – and of protein. A CSF protein level approaching or greater than 100 mg / 100 ml should raise serious doubts as regards a diagnosis of MS.

Measurement of the level of IgG is an important part of the CSF examination. Despite the introduction and widespread use of several formulas for measuring the rate of production of intrathecal IgG, the simplest and most reliable measurement is the percentage of total protein: >15% is considered abnormal. However, an elevated CSF IgG is non-specific. It is often observed in encephalomyelitis, Guillain–Barré syndrome, neurosyphilis, HTLV-I-associated paraparesis, Lyme disease, sarcoidosis and many other infections as well as being a remote effect of cancer (Figure 56).

Another important aspect of CSF examination in MS is the search for oligoclonal bands in the gammaglobulin fraction of protein. To be significant, there must be at least two oligoclonal bands present and none should be found in the serum (Figure 57). The presence of oligoclonal bands in the CSF is not specific for MS; they may be noted in all of the conditions already mentioned above for IgG. This is particularly true of the various forms of disseminated encephalomyelitis, but a valuable differential feature is that the bands, which never disappear from the CSF in MS, may do so in disseminated encephalomyelitis.

Evoked potential studies

Pattern-reversal visual evoked responses are particularly useful in identifying optic nerve and chiasmatic lesions in patients who have had no symptoms or signs of involvement of the visual system. These responses may be positive in up to 75% of such patients, including those with normal visual acuity. The critical measurement is that of the peak, designated as P 100. The amplitude of the response is of little portent. Interocular differences in P 100 delay are most often due to differences in visual acuity between the two eyes and are meaningless. Because the response is modified by changes in visual acuity, it is imperative that the patient wear prescribed corrective lenses during the test. Delay of P 100 is far from specific for MS lesions of the optic system; in addition to poor fixation and changes in visual acuity, many other conditions may give false-positive results. Among the more common ones are glaucoma, alcohol ingestion, cerebrovascular disease, spinocerebellar degeneration, all types of optic atrophy and use of many commonly prescribed drugs (Figure 58).

Brain-stem auditory responses are much less useful because they are positive in only a small percentage of cases, even in the presence of overt clinical signs of brain-stem involvement, such as internuclear ophthalmoplegia. This test has the advantage of not requiring the patient's cooperation. The response is divided into the following waveforms:

Wave I, which originates in the auditory nerve;

Wave II, which originates in the cochlear nucleus;

Wave III, which originates in the superior olivary complex; and

Waves IV and V, thought to originate from the region of the lateral lemniscus and inferior colliculus.

In patients with MS, the most frequently observed delay is between waves III and V (Figure 59). Peripheral delays simply indicate acoustic nerve lesions, which are usually unrelated to MS. Delays between waves I and III may be seen, indicating a lesion in or near the cochlear nuclei.

Somatosensory evoked responses are usually not useful and, indeed, should rarely be performed. These responses are, however, obtained most frequently in patients who have clinical evidence of spinal cord involvement requiring no confirmation. They are often difficult to obtain, especially in the lower extremities, and may be uncomfortable or even painful for the patient.

Neuroimaging

Attempts to visualize the lesions of MS date back to the 1950s, when radioactive phosphorus was used, but this technique could only reveal very large plaques (Figure 60). The late 1970s saw the advent of computed tomography (CT), an important technological breakthrough, which was followed a few years later by the enormously successful MRI.

Computed tomography (CT)

CT scanning, however, will undoubtedly remain, for many years to come, the only neuroimaging procedure generally available in the poorer countries. Although it is true that CT resolution is far below that of MRI, the cost of the equipment and the scan itself are only a fraction of the cost of MRI. The introduction of intravenous iodinated contrast medium capable of penetrating the partially permeable BBB has increased the usefulness of CT. Doubling, or even tripling, the dose of contrast and delaying the imaging process for one or two hours after contrast injection have also greatly enhanced the ability of CT to reveal MS lesions, even in the spinal cord (Figures 61–73).

Magnetic resonance imaging (MRI)

The introduction of MRI, on the other hand, has completely revolutionized the diagnostic process of MS, but has proved, in fact, to be a mixed blessing. The proliferation of MRI machines in the more affluent countries has led to their overuse and to misinterpretation of the images. At present, too often the diagnosis of MS has been based exclusively on the presence of 'lesions' visualized in the white matter on T_2-weighted MRI scans.

A significant number (approximately 5–15%) of clinically definite MS patients have completely normal MRIs on repeated examination. Conversely, there are patients with a variety of insignificant complaints who have MRI abnormalities that are similar to those frequently seen in symptomatic MS patients. The correlation between the number, site and size of MRI white-matter 'lesions', and the clinical signs and symptoms of MS is unreliable. There is no absolutely certain information regarding the nature of the changes seen on MRI that may be considered to represent true MS lesions; the often-used term 'burden of disease' is misleading, as truly enormous areas of presumed demyelination may be seen which have persisted for years in clinically normal subjects.

Attempts to establish reliable MRI diagnostic criteria have failed because the pattern and characteristics of images associated with MS are also seen in many other diseases. There are no MRI patterns of 'lesions', including the ovoid periventricular lesion, which are essential or even diagnostic of MS. Despite this fact, an increasing number of patients are erroneously diagnosed as having MS, and some are even started on specific prophylactic drug treatment based exclusively on MRI interpretation by a general radiologist with little, if any, clinical information. Furthermore, many neurologists do not have the opportunity to review the MRI films first-hand (Figures 74–86).

A very important, but rarely emphasized, use of MRI is in the routine visualization of the cervical cord. In a surprisingly large number of MS patients, cervical cord plaques can be ascribed to compression – whether actual, potential or intermittent – by spondylosis and / or herniated disks (Figures 87–91). Such extrinsic lesions may cause aggravation and progression of the underlying MS in addition to the myelopathic effects they produce. Obtaining lateral MRI views of the neck in flexion, when possible, may reveal an increased effacement of the ventral subarachnoid space or even cord compression that is not evident with the neck in a normal position. Such a lesion should be particularly suspected in patients who have suffered a whiplash or other neck injury (Figure 92).

Gadolinium enhancement of MRI has limited clinical application. It is useful only in cases where it is important to determine the age of the lesions, such as in reference to recent potential triggering events. It is unusual to find an enhanced lesion that is not visible on the T_2-weighted image. Treatment decisions of exacerbations should be based on clinical considerations. At present, the reliability of evaluating long-term treatment by serial gadolinium-enhanced MRI has not been completely settled.

Imaging differential diagnoses

Many diseases of the nervous system that result in white-matter lesions seen by MRI are often erroneously diagnosed as MS. By far, the most common of these is acute disseminated encephalomyelitis (ADEM), which is caused by an infection almost always due to a virus or vaccination. However, the distribution and especially the size of the lesions in ADEM are different and should only rarely be confused with MS (Figures 93–101).

More difficult is the clinical differentiation of the rarer types of disseminated encephalomyelitis – the chronic, the recurrent (RDEM), and the multiphasic (MDEM). The MRI changes are of the same type as those seen in ADEM. However, by definition, ADEM is a monophasic event whereas RDEM is characterized by symptomatic stereotypy. Following the initial bout of ADEM, one or more episodes may occur that reproduce some or all of the original symptoms. In MDEM, separate and clinically different events may occur after the initial episode of ADEM. The MDEM clinical history is identical to that of MS, but the MRI appearances should establish the correct diagnosis (Figure 102).

Other diseases which are frequently mistaken for MS on the basis of their MRI appearances include Lyme disease, HTLV-I-associated paraparesis, AIDS, cerebral arteritis such as lupus erythematosus, complicated migraine, trauma, cerebrovascular disease, neurosarcoidosis and the chronic fatigue syndrome (CFS). This syndrome, occurring mostly in young women, shares with MS the same characteristic fatigue as well as some of the more typical symptoms and signs. In contrast, the following symptoms are useful in establishing the diagnosis of CFS rather than MS: migratory myalgia, arthralgia and painful paresthesiae, sleep disturbance, anhedonia, and unusual and paradoxical reactions to medication (Figures 103–112).

Single-photon emission computed tomography (SPECT)

This technique uses radioactive tracers with computerized reconstruction of the isotopic emissions recorded by a rotating gamma camera to measure cerebral perfusion. The applicability of SPECT to MS is limited to specialized research (Figures 113 and 114).

Positron emission tomography (PET)

Using radioisotopes with very short half-lives, PET is a method for the quantitative measurement of cerebral metabolism. Radiolabeled deoxyglucose is one of the most commonly studied metabolites; it has been reported to show as much as a 20–30% reduction in oxygen or glucose consumption in various parts of the brain in MS patients. However, PET is still primarily a research tool and requires close proximity to a cyclotron (Figures 115–117).

Conclusion

Despite major gaps in our knowledge of the pathogenesis, epidemiology and genetics of MS, much progress has been made not only in those areas of research, but also in refining confirmatory laboratory diagnostic procedures and improving methods of symptomatic treatment. Clinical research activity in prophylactic therapy continues with some promising, but as yet unproved, results. The issue that remains unchallenged is the primary role played by the astute and experienced clinician in the diagnosis and treatment of the disease.

Bibliography

Adams C, Poston R, Buk S, *et al.* Inflammatory vasculitis in multiple sclerosis. *J Neurol Sci* 1985;69: 269–83

Allen I. Pathology of multiple sclerosis. In: Matthews W, ed. *McAlpine's Multiple Sclerosis*, 2nd edn. Edinburgh: Churchill Livingstone, 1991:341–78

Allen I, Kirk J. Demyelinating diseases. In: Adams J, Duchen L, eds. *Greenfield's Neuropathology*, 5th edn. London: Edward Arnold, 1992:447–520

American Academy of Ophthalmology. Intermediate uveitis. In: *Basic and Clinical Science Course, Section 9: Intraocular Inflammation and Uveitis.* San Francisco: American Academy of Ophthalmology, 1997–1998:102

Barnes D, Munro P, Youl B, *et al.* The long-standing MS lesion. *Brain* 1991;114:1271–80

Brain R, Wilkinson M. The association of cervical spondylosis and disseminated sclerosis. *Brain* 1957; 80:456–78

Brody J, Sever J, Henson T. Virus antibody titers in multiple sclerosis patients, siblings and controls. *J Am Med Assoc* 1971;216:1441–6

Broman T. Supravital analysis of disorders in the cerebral vascular permeability. II. Two cases of multiple sclerosis. *Acta Psychiatr Neurol Scand* 1947;46(suppl):58–71

Cannella B, Raine C. The adhesion molecule and cytokine profile of multiple sclerosis lesions. *Ann Neurol* 1995;37:424–35

Chen C-J, Ro L-S, Chang C-N, *et al.* Serial MRI studies in pathologically verified Balo's concentric sclerosis. *J Comput Assist Tomogr* 1996;20: 732–5

Dawson J. The histology of disseminated sclerosis. *Trans R Soc Edinburgh* 1916;50:517–740

Dean G, Kurtzke J. On the risk of acquiring multiple sclerosis according to age at immigration to South Africa. *Br Med J* 1971;3:725–9

Ebers G, Kukay K, Bulman D, *et al.* A full genome search in multiple sclerosis. *Nat Genet* 1996; 13:472–6

Fazekas F, Offenbacher H, Fuchs S, *et al.* Criteria for an increased specificity of MRI interpretation in elderly subjects with suspected multiple sclerosis. *Neurology* 1988;38:1822–5

Firth D. *The Case of Augustus d'Esté*. Cambridge: Cambridge University Press, 1948

Fog T. The topography of plaques in multiple sclerosis. *Acta Neurol Scand* 1965;41(suppl 15): 1–161

Gay F, Drye T, Dick G, *et al.* The application of multifactorial cluster analysis in the staging of plaques in early multiple sclerosis. Identification and characterization of the primary demyelinating lesion. *Brain* 1997;120:1461–83

Gay F, Esiri M. Blood–brain barrier damage in acute multiple sclerosis plaques. *Brain* 1991;114: 557–72

Gharagozloo A, Poe L, Collins G. Antemortem diagnosis of Baló's concentric sclerosis: Correlative MR imaging and pathologic findings. *Radiology* 1994;191:817–9

Gonsette R, André-Balisaux G, Delmotte P. La perméabilité des vaisseaux cérébraux. IV. Démyélinisation expérimentale provoquée par des substances agissant sur la barrière hématoencéphalique. *Acta Neurol Belg* 1966;66:247–62

Haegert D, Swift F, Benedikz J. Evidence for a complex role of HLA class II genotypes in susceptibility to multiple sclerosis in Iceland. *Neurology* 196;46:1107–11

Hanemann C, Kleinschmidt A, Reiffenberger G, *et al.* Baló's concentric sclerosis followed by MRI and positron emission tomography. *Neuroradiology* 1993;35:578–80

Hickey W. Migration of hematogenous cells through the blood–brain barrier and the initiation of CNS inflammation. *Brain Pathol* 1991; 1:97–105

Jahnke U, Fischer E, Alvord E. Sequence homology between certain viral proteins and proteins related to encephalomyelitis and neuritis. *Science* 1985;229:282–4

Jansen H, Willemsen A, Sinnige L, *et al.* Cobalt-55 positron emission tomography in relapsing–progressive multiple sclerosis. *J Neurol Sci* 1995; 132:139–45

Johnson M, Lavin P, Whetsell W Jr. Fulminant monophasic multiple sclerosis, Marburg's type. *J Neurol Neurosurg Psychiatr* 1990;53:918–21

Keltner J, Johnson C, Spurr J, *et al.* Baseline visual field profile of optic neuritis. *Arch Ophthalmol* 1993;111:231–4

Khan S, Yaqub B, Poser C, *et al.* Multiphasic encephalomyelitis presenting as alternating hemiplegia. *J Neurol Neurosurg Psychiatr* 1995; 58:467–70

Kinnunen E, Valle M, Piiranen L, *et al.* Viral antibodies in MS: A nationwide co-twin study. *Arch Neurol* 1990;47:743–6

Kirk J. The fine structure of the CNS in multiple sclerosis. II. Vesicular demyelination in an acute case. *Neuropathol Appl Neurobiol* 1979;5:289–94

Kurtzke J, Gudmundsson K, Bergmann S. Multiple sclerosis in Iceland. I. Evidence of a postwar epidemic. *Neurology* 1982;32:143–50

Kurtzke J, Hyllested J. Multiple sclerosis in the Faroe Islands. I. Clinical and epidemiological features. *Ann Neurol* 1979;5:6–21

Lightman S, McDonald W, Bird A, *et al.* Retinal venous sheathing in optic neuritis: Its significance for the pathogenesis of multiple sclerosis. *Brain* 1987;110:405–14

Lycke J, Wikkelsö C, Bergh A-C, *et al.* Regional cerebral blood flow in multiple sclerosis measured by single-photon emission tomography with technetium-99m hexamethylpropyleneamine oxime. *Eur Neurol* 1993;33:163–7

McFarland H, Greenstein J, McFarlin D, *et al.* Family and twin studies in multiple sclerosis. *Ann NY Acad Sci* 1984;436:118–24

McFarland H, Patronas N, McFarlin D, *et al.* Studies in multiple sclerosis in twins using magnetic nuclear resonance. *Neurology* 1985;35(suppl): 137

Medaer R. Does the history of multiple sclerosis go as far back as the 14th century? *Acta Neurol Scand* 1979;60:188–92

Namerow N, Thompson L. Plaques, symptoms and the remitting course of multiple sclerosis. *Neurology* 1969;19:765–74

Olsson T, Diener P, Ljungdahl A, *et al.* Facial nerve transection causes expansion of myelin autoreactive T cells in regional lymph nodes and T cell homing to the facial nucleus. *Autoimmunity* 1992;13:117–26

Olsson T, Sun J, Solders G, *et al.* Autoreactive T and B cell responses to myelin antigens after diagnostic sural nerve biopsy. *J Neurol Sci* 1993; 117:130–9

Oppenheimer D. The cervical cord in multiple sclerosis. *Neuropathol Appl Neurobiol* 1978;4: 151–62

Orrell R, Shakir R, Lane R, *et al.* Distinguishing acute disseminated encephalomyelitis from multiple sclerosis. *Br Med J* 1996;313:802–4

Ozawa K, Suchanek S, Breitschopf H, *et al.* Patterns of oligodendroglial pathology in multiple sclerosis. *Brain* 1994;117:1311–22

Paty D, Poser C. Clinical symptoms and signs of multiple sclerosis. In: Poser C, Paty D, Scheinberg L, *et al.*, eds. *The Diagnosis of Multiple Sclerosis.* New York: Thieme–Stratton, 1984: 27–43

Poser C. Exacerbations, activity and progression in multiple sclerosis. *Arch Neurol* 1980;37:471–4

Poser C. The course of multiple sclerosis. *Arch Neurol* 1985;42:1035

Poser C. The pathogenesis of multiple sclerosis. A critical reappraisal. *Acta Neuropathol* 1986;71: 1–10

Poser C. The peripheral nervous system in multiple sclerosis: A review and pathogenetic hypothesis. *J Neurol Sci* 1987;79:83–90

Poser C. Magnetic resonance imaging in asymptomatic disseminated vasculomyelinopathy. *J Neurol Sci* 1989;94:69–77

Poser C. Multiple sclerosis. Observations and reflections – a personal memoir. *J Neurol Sci* 1992;107:127–40

Poser C. The pathogenesis of multiple sclerosis. Additional considerations. *J Neurol Sci* 1993;115 (suppl):S3–15

Poser C. The epidemiology of multiple sclerosis. A general overview. *Ann Neurol* 1994a;36(S2): S180–93

Poser C. The pathogenesis of multiple sclerosis. *Clin Neurosci* 1994b;2:258–65

Poser C. The role of trauma in the pathogenesis of multiple sclerosis. A review. *Clin Neurol Neurosurg* 1994c;96:103–10

Poser C. Myalgic encephalomyelitis / chronic fatigue syndrome and multiple sclerosis: Differential diagnosis. *EOS J Immunol Immunopharmacol* 1995a;196:765–71

Poser C. Onset symptoms of multiple sclerosis. *J Neurol Neurosurg Psychiatr* 1995b;58:253–4

Poser C, Goutières F, Carpentier M-A, *et al.* Schilder's myelinoclastic diffuse sclerosis. *Pediatrics* 1986;77:107–12

Poser C, Hibberd P, Benedikz J, *et al.* An analysis of the "epidemic" of multiple sclerosis in the Faroe Islands. *Neuroepidemiology* 1988;7:168–80

Poser C, Kleefield J, O'Reilly G, *et al.* Neuroimaging and the lesion of multiple sclerosis. *AJNR* 1987; 8:549–52

Poser C, Paty D, Scheinberg L, *et al.* New diagnostic criteria for multiple sclerosis. *Ann Neurol* 1983;13:227–31

Poser C, Roman G, Vernant J-C. Multiple sclerosis or HTLV-I myelitis? *Neurology* 1990;40:1020–2

Pozzilli C, Bernardi S, Mansi U, *et al.* Quantitative assessment of blood–brain barrier permeability in multiple sclerosis using 68-Ga-EDTA and positron emission tomography. *J Neurol Neurosurg Psychiatr* 1988;51:1058–62

Raine C. The immunology of multiple sclerosis. *Ann Neurol* 1994;36:S61–72

Robertson N, Clayton D, Fraser M, *et al.* Clinical concordance in sibling pairs with multiple sclerosis. *Neurology* 1996;47:347–52

Roelcke U, Kappos L, Lechner-Scott J, *et al.* Reduced glucose metabolism in the frontal cortex and basal ganglia of multiple sclerosis patients with fatigue. *Neurology* 1997;48:1566–71

Roizin L, Helfand M, Moore J. Disseminated diffuse and transitional demyelination of the central nervous system. *J Nerv Ment Dis* 1946;104:1–50

Sadovnick A, Ebers G, Dyment D, *et al.* Evidence for genetic basis of multiple sclerosis. *Lancet* 1996;347:1728–30

Sagar H, Warlow C, Sheldon P, *et al.* Multiple sclerosis with clinical and radiological features of cerebral tumour. *J Neurol Neurosurg Psychiatr* 1982;45:802–8

Thomas P, Walker R, Rudge P, *et al.* Chronic demyelinating peripheral neuropathy associated with multifocal central nervous system demyelination. *Brain* 1987;110:53–76

Tsai M-L, Hung K-L. Multiphasic disseminated encephalomyelitis mimicking multiple sclerosis. *Brain Dev* 1996;18:412–4

Woyciechowska J, Dambrozia J, Chu A, *et al.* Correlation of oligoclonal IgG bands and viral antibodies in twins with multiple sclerosis. In: Cazzullo C, Caputo D, Ghezze A, *et al.*, eds. *Virology and Immunology in Multiple Sclerosis.* Berlin: Springer-Verlag, 1987:45–9

Woyciechowska J, Dambrozia J, Leinikki P, *et al.* Viral antibodies in twins with multiple sclerosis. *Neurology* 1985;35:1176–80

Section 2 Multiple Sclerosis Illustrated

List of illustrations

Figure 20
Higher power view of brain section showing new MS plaque next to a vascular infarct

Figure 21
Delayed contrast axial CT scans of severe MS and MRIs taken after recovery

Figure 22
Schematic diagram of the Hawaiian Islands

Figure 23
Unstained brain section showing typical periventricular and parenchymatous lesions

Figure 24
Unstained brain section showing typical periventricular and parenchymatous lesions

Figure 25
Unstained brain section showing typical periventricular and parenchymatous lesions

Figure 26
Unstained section of cerebellum showing typical lesion of MS

Figure 27
Unstained sections of spinal cord showing scattered lesions at all levels

Figure 28
Celloidin sections of occipital lobe (Nissl and Weigert stains)

Figure 29
Celloidin section of brain showing 'shadow' lesions (Weigert stain)

Figure 30
Biopsy section showing lesion with the characteristic well-defined border (Klüver–Luxol fast blue stain)

Figure 31
Celloidin section showing typical sharply defined edges of the lesions (Weigert stain)

Figure 32
Sections through optic chiasm showing optic nerves (Weigert and H & E stains)

Figure 33
Sections of pons and cervical spinal cord showing MS plaques (Weigert stains)

Figure 34
Histological section of edematous parenchymal blood vessel in acute MS (H & E stain)

Figure 35
Histological section of parenchymal blood vessel showing extensive inflammatory changes due to acute MS (H & E stain)

Figure 36
Celloidin section showing age of lesion by presence of abbau products of myelinoclasia (myelin–fat stain)

Figure 37
High-power view of myelin sheaths showing fat-laden macrophages (myelin–fat stain)

Figure 38
Low-power view of fat-laden macrophages in acute MS (oil red–O stain)

Figure 39
Histological views of early MS plaques (H & E–Luxol fast blue and Golgi stains)

Figure 40
Biopsy of cerebral plaque erroneously diagnosed as astrocytoma (PAS stain)

Figure 41
Biopsy of deep thalamic plaque erroneously diagnosed as brain tumor (Bodian and PTAH stains)

Figure 42
Histological section of medulla showing large MS lesion (Luxol fast blue–PAS stain)

Figure 43
Histological section of cervical spinal cord showing large MS lesion (Luxol fast blue–PAS stain)

Figure 44
Sections of spinal cord, medulla and pons showing disseminated lesions of MS (Weigert stains)

Figure 45
Brain sections showing 'onion-bulb' enlargement of spinal and peripheral nerves

Figure 75
T_1- and T_2-weighted, proton-density and gadolinium–EDTA MRIs in histologically definite MS

Figure 76
T_2-weighted MRIs showing bilateral areas of increased signal intensity in definite MS

Figure 77
T_2-weighted and proton-density MRIs in MS patient with severe dementia

Figure 78
T_2-weighted and gadolinium–EDTA MRIs showing multiple lesions

Figure 79
Sagittal and coronal T_1-weighted MRIs of MS patient with prolactinemia

Figure 80
T_2-weighted MRI showing areas of increased signal intensity in the optic tract and superior angles of the ventricles

Figure 81
Coronal and axial T_2-weighted MRIs in a patient with optic neuritis

Figure 82
MRI of brain stem showing areas of increased signal intensity in patient with facial palsy

Figure 83
Audiogram and evoked auditory response and MRI of brain stem of a patient with sudden partial loss of hearing

Figure 84
T_1-weighted MRIs showing hypodense area in the middle cerebellar peduncle

Figure 85
Transverse, coronal and sagittal MRIs of a patient with Baló's disease

Figure 86
T_2-weighted MRIs of the brain and spinal cord of an asymptomatic patient

Figure 87
Sagittal and axial MRIs of spinal cord showing MS plaque

Figure 88
Proton-density sagittal MRI of spinal cord showing disk and cord compression

Figure 89
Sagittal and axial T_2-weighted MRIs showing herniated disks and spondylosis

Figure 90
Axial and sagittal T_2-weighted MRIs showing severe spinal cord spondylosis

Figure 91
Axial and sagittal T_2-weighted MRIs showing cord compression and MS plaques

Figure 92
Sagittal T_2-weighted MRIs of cervical spinal cord post-whiplash injury

Figure 93
T_2-weighted MRIs of acute disseminated encephalomyelitis which appears to fulfill MS MRI criteria

Figure 94
Proton-density MRIs of acute disseminated encephalomyelitis

Figure 95
Proton-density MRIs of acute disseminated encephalomyelitis misdiagnosed as MS

Figure 96
T_2-weighted MRIs of acute disseminated encephalomyelitis with areas of increased signal intensity restricted to the cerebellum

Figure 97
Proton-density MRIs of acute disseminated encephalomyelitis misdiagnosed as Schilder's disease

Figure 98
Sagittal and axial T_2-weighted MRIs of cervical spinal cord in acute disseminated encephalomyelitis

Figure 99
T_2-weighted MRIs of a patient with post-vaccination (influenza) encephalomyelitis

Figure 100
MRIs of acute disseminated encephalomyelitis misdiagnosed as MS

Figure 1 Augustus d'Esté (1794–1848), the first documented case of multiple sclerosis, was reported by Douglas Firth in 1948. Reproduced with the permission of the Victoria and Albert Museum, London

Figure 2 Sir Robert Carswell (1793–1857), Professor of Morbid Anatomy at University College, London, personally painted 2000 water-color illustrations of pathological specimens. He published an atlas in 1838. Courtesy of Countway Library, Harvard Medical School

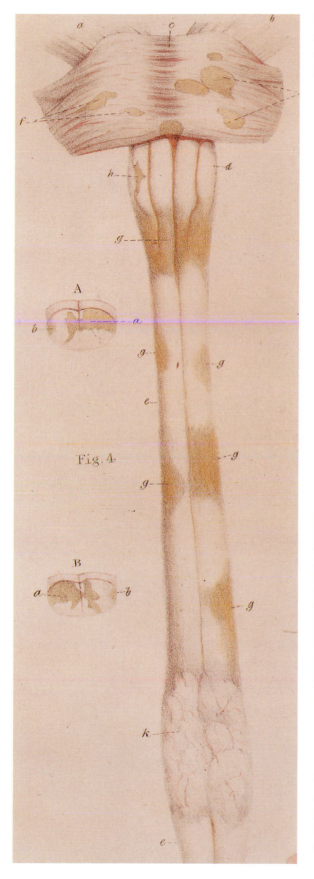

Figure 3 A watercolor of multiple sclerosis of the spinal cord, taken from Carswell's atlas. The plaques were simply called 'atrophy'. Courtesy of Countway Library, Harvard Medical School

Figure 4 Jean Cruveilhier (1791–1874), Professor of Anatomy, and later of Pathology, at the University of Paris. His atlas was published as a series of fascicles between 1829 and 1842. Courtesy of Countway Library, Harvard Medical School

Figure 5 Illustrations of multiple sclerosis of the (a) brain, (b) spinal cord and (c) brain stem, from Cruveilhier's atlas. The plaques were described as "gray transformation and degeneration". Courtesy of Countway Library, Harvard Medical School

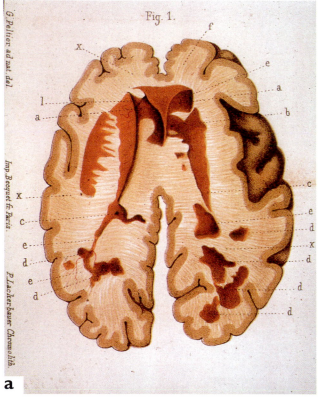

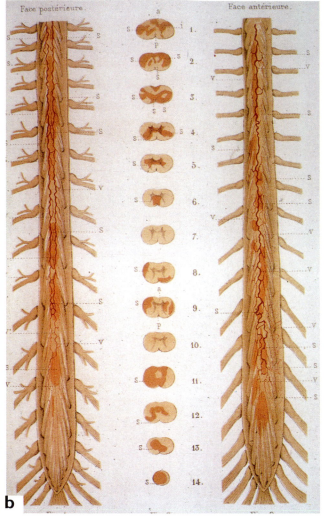

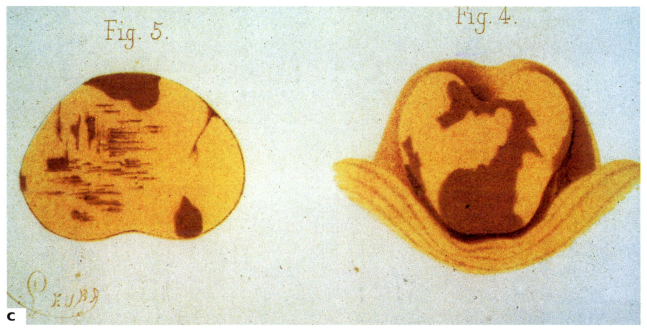

James Walker Dawson

Figure 6 Jean Martin Charcot (1825–1893), Professor of Pathologic Anatomy at the University of Paris and neurologist at the Salpêtrière. Courtesy of Countway Library, Harvard Medical School

Figure 7 James Walker Dawson (1870–1927), Pathologist at the Royal College of Physicians Laboratory in Edinburgh. His monograph, *The Histology of Disseminated Sclerosis*, was published in 1916. Courtesy of Countway Library, Harvard Medical School

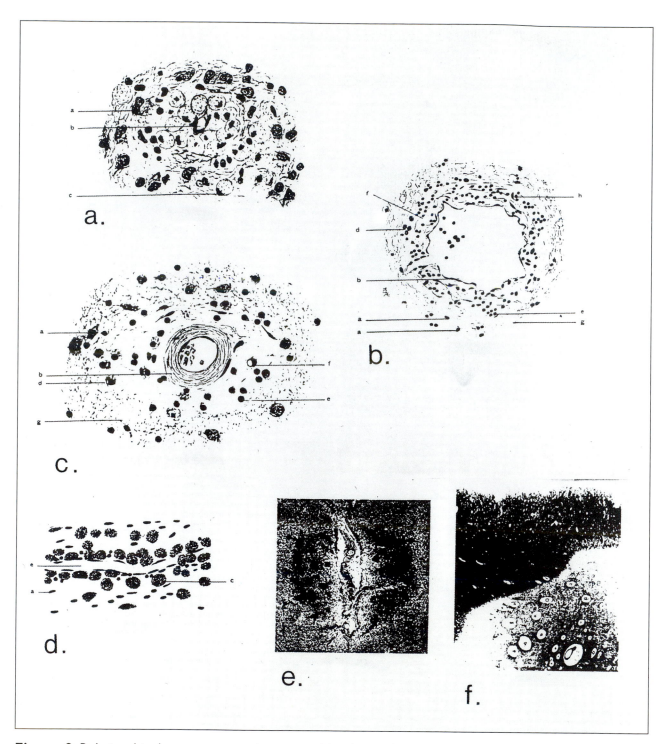

Figure 8 Relationship between parenchymatous blood vessels and lesions of multiple sclerosis, taken from Dawson's monograph (reprinted in 1973 by the Montreal Neurological Institute). (a) Vessel adventitia filled with fat granule cells; there are numerous similar cells and enlarged glia in the surrounding tissue. (b) Cellular infiltration of the adventitia in an area of sclerosis after removal of fat cells. (c) Small artery in an area of advanced sclerosis; hyaline fusion of the media and adventitia is just beginning. (d) Capillary in a demyelinated area surrounded by a layer of fat cells. (e) Fat cells surrounding a vessel in an area of sclerosis. (f) Group of vessels in a sclerotic area with extremely dilated perivascular spaces

Figure 9 Töre Broman, Professor of Neurology, University of Göteborg, Sweden, was the first to demonstrate the alteration of the blood–brain barrier in a multiple sclerosis plaque

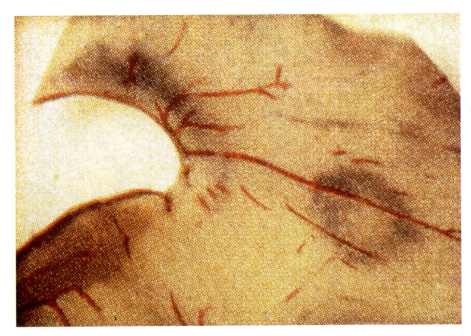

Figure 10 Post-mortem perfusion study of a brain from a patient with multiple sclerosis. The plaques are traversed by venules (stained red). The periphery is stained blue, demonstrating increased permeability of the blood–brain barrier. Supravital trypan blue stain. Courtesy of Professor Töre Broman, University of Göteborg, Sweden

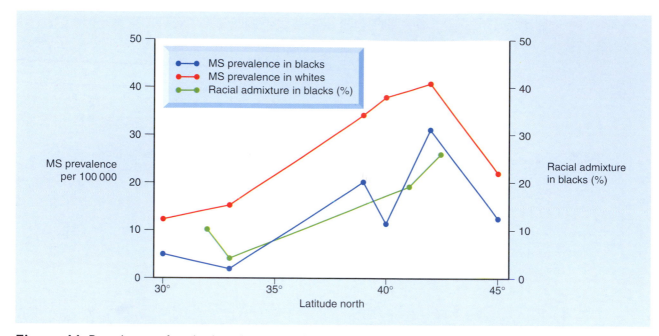

Figure 11 Prevalence of multiple sclerosis in blacks compared with whites in the USA. Note that the percentage of racial admixture in blacks, based on blood group studies, parallels the rate of prevalence in blacks. Modified from Poser, 1992; reproduced with the permission of Elsevier Science BV

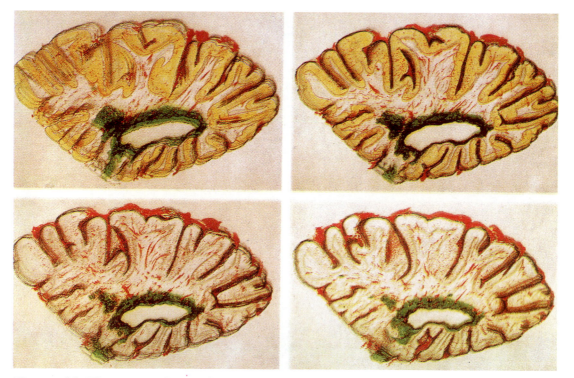

Figure 12 There is a close relationship between parenchymal blood vessels and multiple sclerosis plaques (stained green). From Fog, 1965; reproduced with the permission of Munksgaard International Publishers

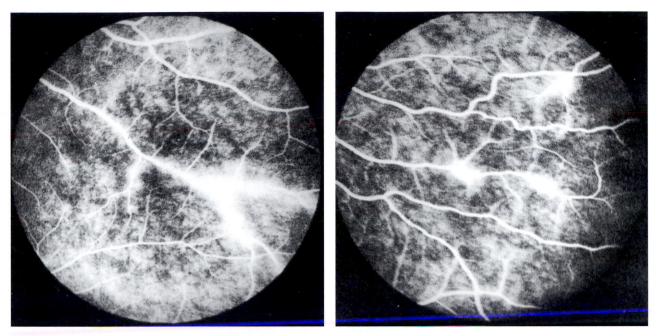

Figure 13 Retinal fluorescein angiography in optic neuritis. Note leakage from vessels, indicating alteration of permeability of the vessel walls in the retina, a myelin-free structure. From Lightman *et al.*, 1987; reproduced with the permission of Oxford University Press

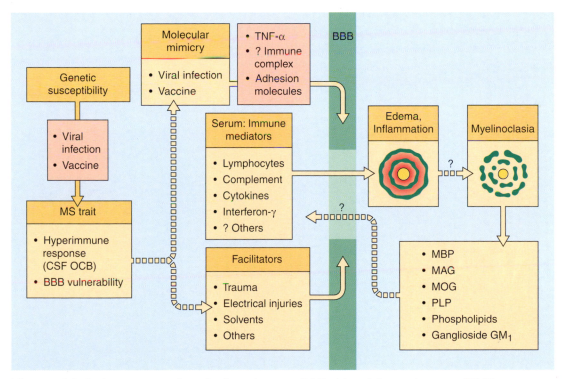

Figure 14 Pathogenesis of multiple sclerosis. BBB, blood–brain barrier; TNF-α, tumor necrosis factor-α; OCB, oliogoclonal bands; MBP, myelin basic protein; MAG, myelin-associated glycoprotein; MOG, myelin-oligodendroglia glycoprotein; PLP, proteolipid protein. Modified from Poser, 1994b; reproduced with the permission of John Wiley & Sons

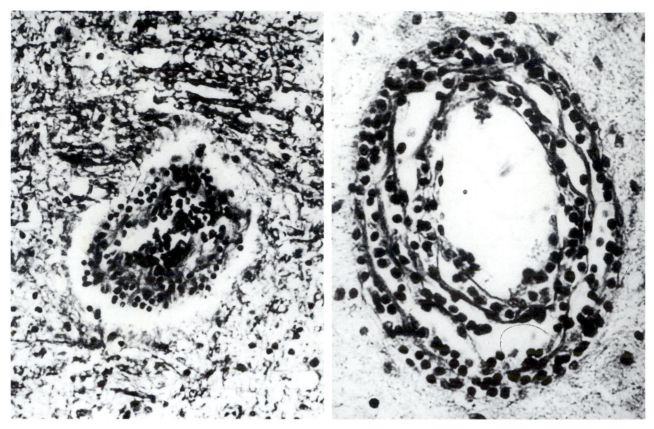

Figure 15 Microscopy showing (left) lymphocytic infiltration confined to the venous wall in normal white matter in active multiple sclerosis (MS), and (right) lymphocytic infiltration and edematous onion-skin changes in the venous wall in normal white matter approximately 1.5 cm from an active MS plaque. From Adams *et al.,* 1985; reproduced with the permission of Elsevier Science BV

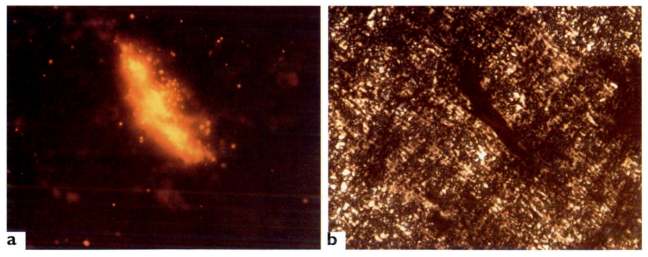

Figure 16 Blood–brain barrier (BBB) defects in normal white matter in multiple sclerosis: (a) Section prepared with rabbit antihuman fibrinogen–tetramethyl rhodamine isothiocyanate (TRITC) fluorescein shows fibrinogen leaking from capillary; (b) same section viewed under polarized light shows normal myelin birefringence except in the vessel walls

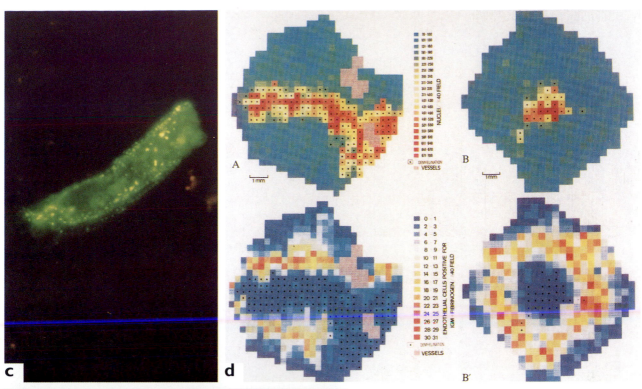

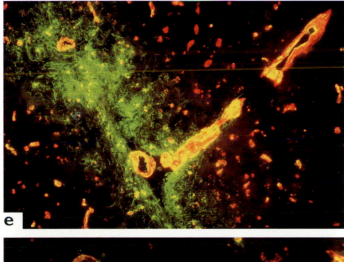

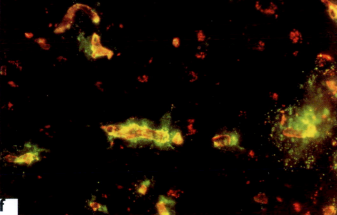

Figure 16 continued
(c) capillary endothelial cells containing vesicles positive for IgM fluorescein isothiocyanate (FITC); (d) distribution of BBB abnormalities associated with two acute plaques. The plaques are defined by their hypercellularity (nuclear counts; upper) and myelin lysis (dotted areas). Both plaques (lower) are surrounded by zones of perivascular fibrinogen leakage (see a) and endothelial cell endocytosis of IgM (see c); (e) blood vessel in normal white matter double-labeled for collagen IV–TRITC and fibrinogen–FITC; and (f) capillaries labeled for collagen IV–TRITC are leaking fibrinogen–FITC into the cortex. From Gay & Esiri, 1991 (a–d); reproduced with the permission of Oxford University Press; and Dr Frederick Gay, Anglia Polytechnic University Cambridge, UK (e & f). For technical details, see Gay *et al.*, 1997

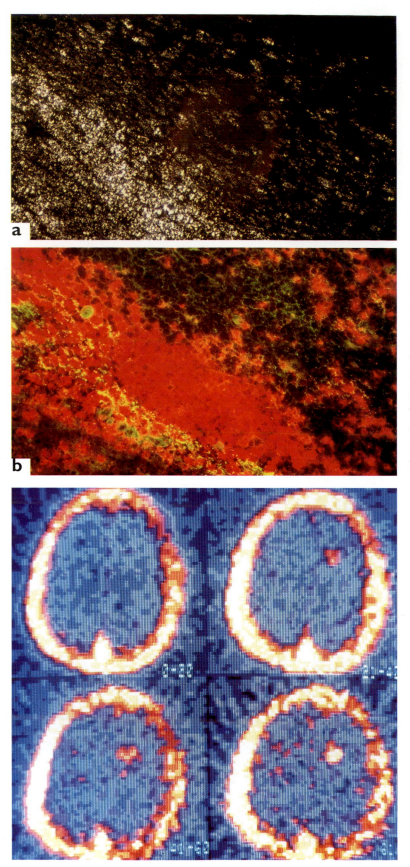

Figure 17 Multiple sclerosis plaques (a) viewed under polarized light, and (b) double-labeled for macrophages (TRITC) and fibrinogen (FITC), and viewed with combined TRITC–FITC blue/green fluorescence filters. A high macrophage density is associated with fibrinogen leakage (yellow). Courtesy of Dr Frederick Gay, Anglia Polytechnic University, Cambridge, UK. For technical details, see Gay et al., 1997

Figure 18 Increased permeability of the blood–brain barrier in a multiple sclerosis plaque: Sequential scans obtained at 20, 40, 60 and 80 min (counterclockwise) after intravenous injection of gadolinium-68–EDTA. Focal areas of increased tracer uptake are shown both in the right (early scan) and left (tardive scan) hemispheres. Courtesy of Dr Carlo Pozzilli, University of Rome 'La Sapienza'. For technical details, see Pozzilli et al., 1988

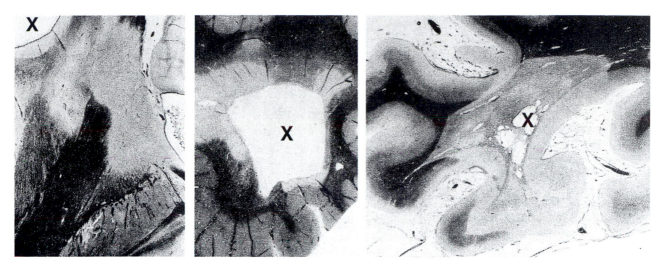

Figure 19 New multiple sclerosis plaques adjacent to brain needle tract (marked by the **X**), the result of gross breach of the blood–brain barrier. Courtesy of Dr Richard Gonsette, Brussels, Belgium

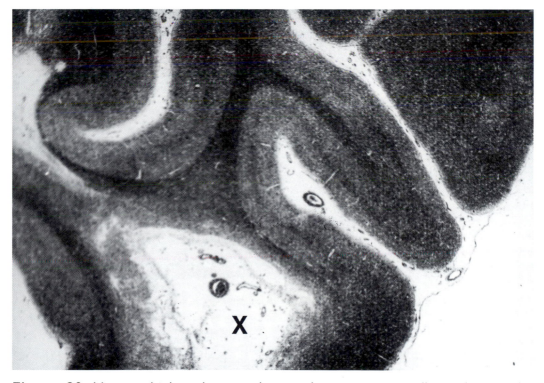

Figure 20 New multiple sclerosis plaque adjacent to a small cerebrovascular accident (marked by the **X**) secondary to increased permeability of the blood–brain barrier. Courtesy of Professor Antonio Nunes Vicente, Coimbra, Portugal

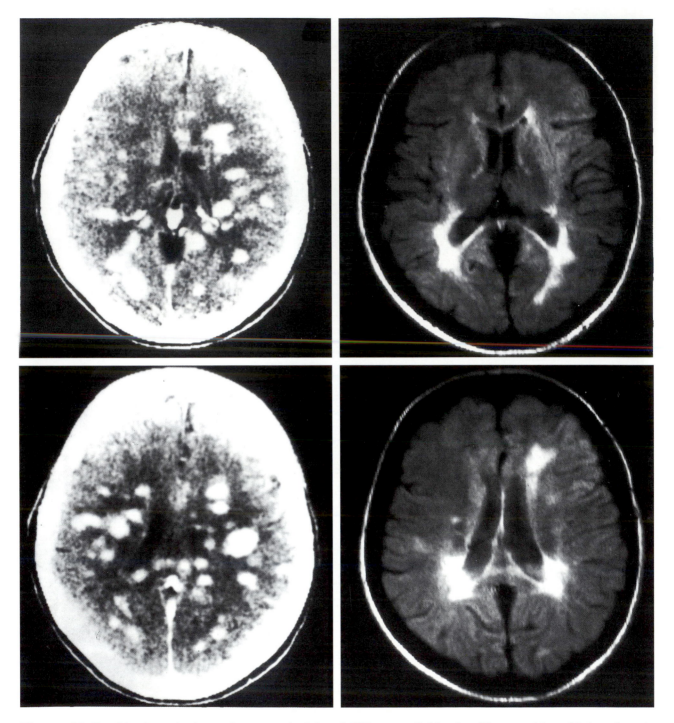

Figure 21 Double-dose (iodinated contrast) delayed CT scans (left) of a 32-year-old multiple sclerosis patient who experienced a severe exacerbation, show numerous areas of enhancement. The MRIs (right), obtained approximately 2 months later, after recovery, show few areas of increased signal intensity (AISI). With the possible exception of the enhanced area at the superior angle of the left ventricle, none of the areas of BBB alteration seen on CT are evident on MRI. Their disappearance so soon after clinical recovery suggests that the contrast enhancement represents edema rather than demyelination, as re-myelination does not occur within such a short period of time. From Poser *et al.*, 1987; reproduced with the permission of the American Society of Radiology

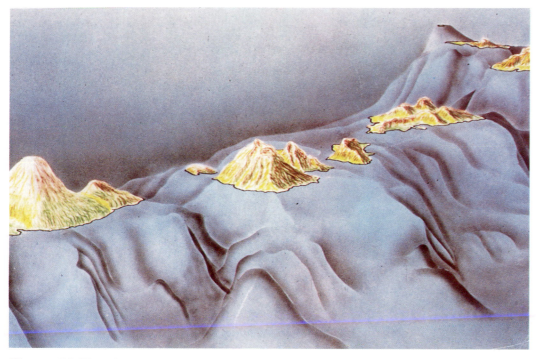

Figure 22 The disease process in multiple sclerosis can be compared with volcanic island chains such as the Hawaiian Islands, where only the tips of the volcanic structures protrude above sea level while unseen activity continues below

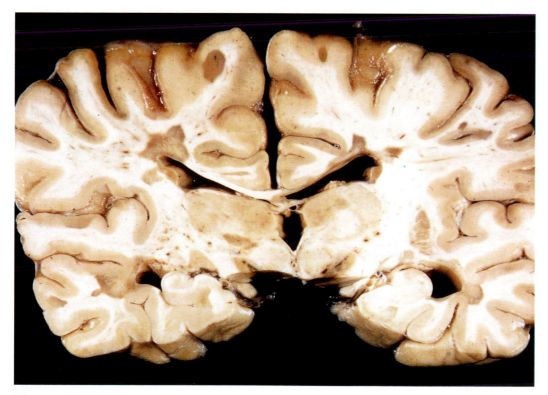

Figure 23 Unstained gross coronal section of brain. Note the typical dark-colored periventricular and parenchymatous lesions

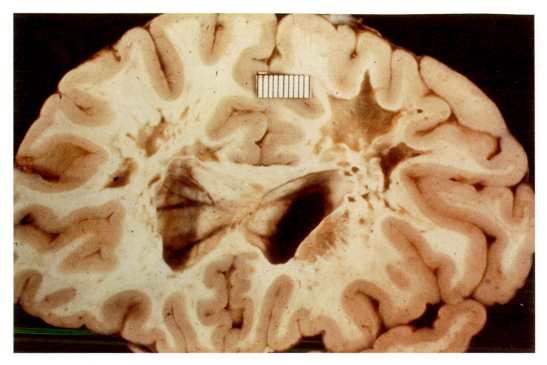

Figure 24 Unstained gross coronal section of brain shows the typical periventricular and parenchymatous lesions, including a large subcortical lesion in the right parietal white matter

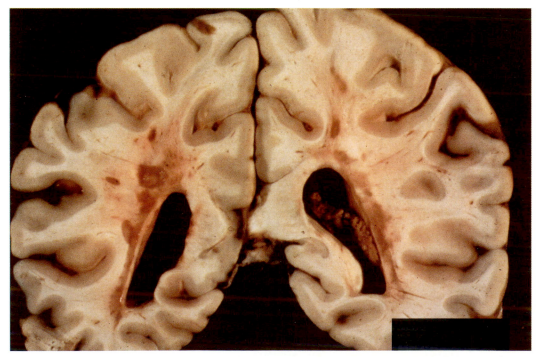

Figure 25 Unstained gross coronal section of brain shows the typical, slightly hemorrhagic, periventricular and parenchymatous lesions. Courtesy of Dr William Pendlebury, University of Vermont, Burlington, VT

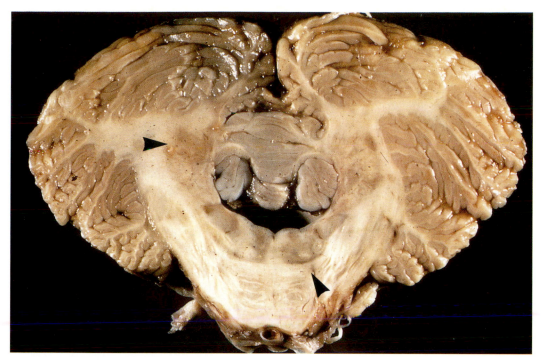

Figure 26 Unstained gross axial section of cerebellum. Note the typical involvement of the dentate nuclei and periventricular lesions (arrows)

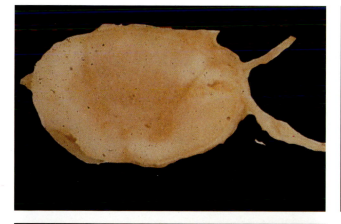

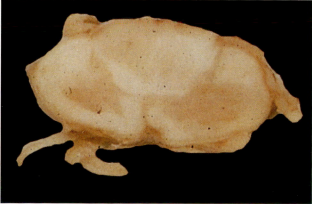

Figure 27 Unstained multiple gross axial sections of spinal cord showing scattered lesions at all levels. Note involvement of the posterior columns at the level of C4 (upper left), and of the anterior horn at C6 (lower left) and pyramidal tract at T6 (upper right)

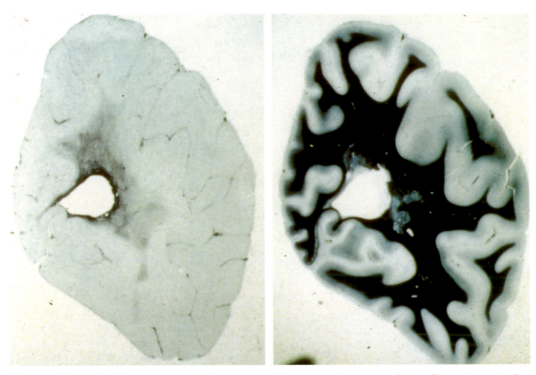

Figure 28 Celloidin sections of occipital lobe showing several small periventricular areas of demyelination. Nissl staining (left) shows reactive gliosis extending well beyond the plaques; Weigert staining (right) shows pale-staining, relatively restricted, demyelination. Courtesy of Professor H. Shiraki, University of Tokyo, Japan

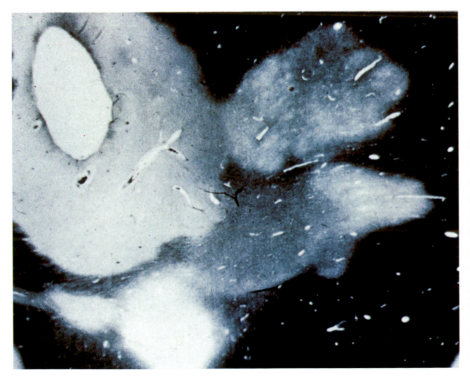

Figure 29 Celloidin section showing different intensities of staining. There is total loss of myelin in some plaques, represented by the absence of color whereas, in other plaques, there is only pallor of the myelin. These 'shadow' plaques are believed by some to represent remyelination. Weigert stain. Courtesy of Professor H. Shiraki, University of Tokyo, Japan

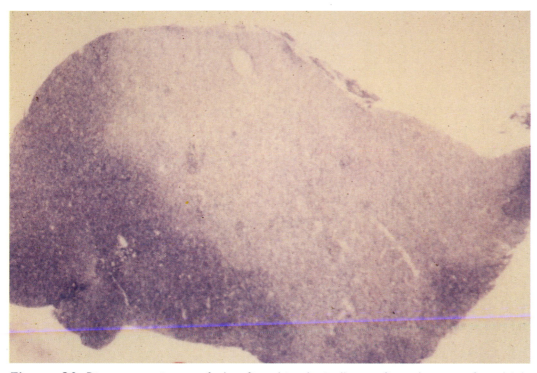

Figure 30 Biopsy specimen of the first histologically confirmed case of multiple sclerosis in a black South African. Note the characteristic clear-cut edge of the lesion. Klüver–Luxol fast blue stain. (Figure 75 shows the MRI of this patient.) Courtesy of Dr S. Nielsen, University of the Witwatersrand, Johannesburg, South Africa

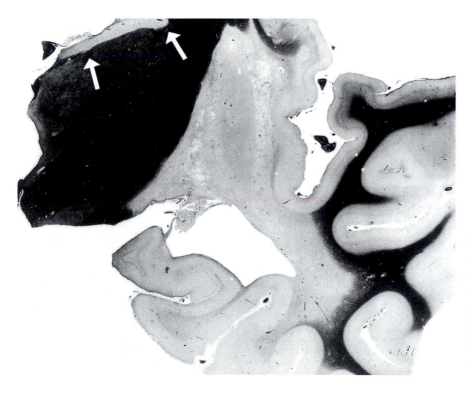

Figure 31 Celloidin section shows the typical well-defined edges of multiple sclerosis lesions with Weigert staining. A periventricular lesion is also present (arrows). Courtesy of Professor H. Shiraki, University of Tokyo, Japan

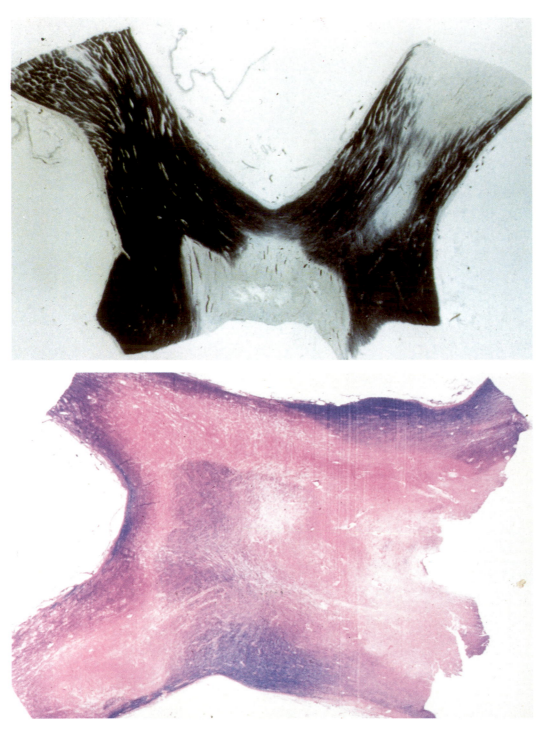

Figure 32 Sections of optic chiasm and optic nerves showing the well-defined edges of a multiple sclerosis lesion in the chiasm and a large lesion in the right optic nerve (upper; Weigert stain), and several distinct plaques in the chiasm and left optic nerve (lower; H & E stain). Courtesy of Dr William Pendlebury, University of Vermont, Burlington, VT

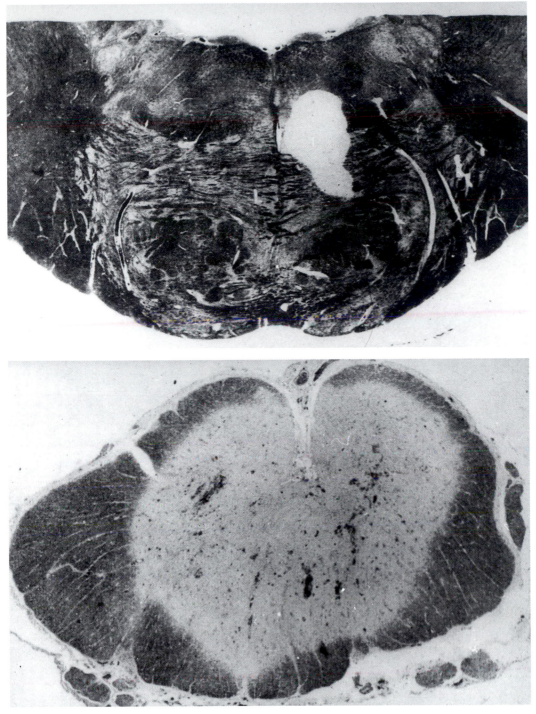

Figure 33 Weigert staining reveals a well-delineated multiple sclerosis plaque in the pons (upper) and a large plaque involving nearly the entire section of the cervical spinal cord (lower). The patient was a black Senegalese. Courtesy of Professor Michel Dumas, Institute of Neurological Epidemiology and Tropical Neurology, Limoges, France

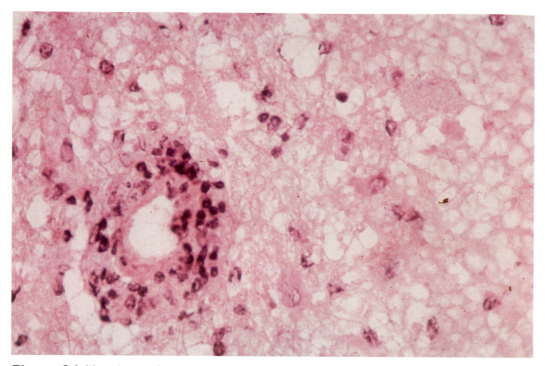

Figure 34 Histology of a plaque in acute multiple sclerosis shows an inflammatory reaction in the wall of a markedly edematous blood vessel. Severe edema of the surrounding parenchyma has produced a spongy appearance. H & E stain

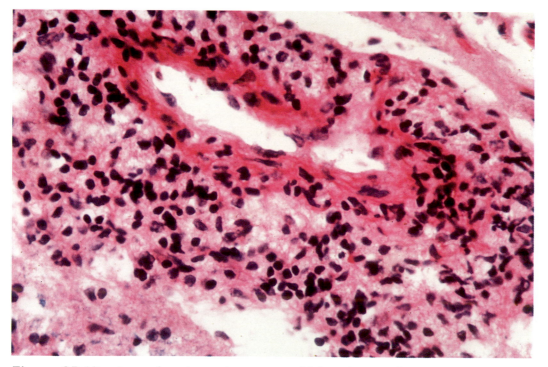

Figure 35 Histology of a plaque in acute multiple sclerosis shows extensive tissue disorganization with an inflammatory reaction and edema of the blood vessel wall and perivascular tissue. H & E stain

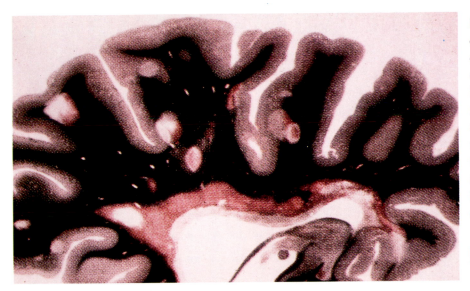

Figure 36 Celloidin section of brain shows that the age of multiple sclerosis lesions can be gauged from the color intensity of the stained fat, representing the abbau products of myelinoclasia: the redder the fat, the younger the lesion. Abbau products are no longer present in old lesions. Combined myelin–fat stains. From Roizin *et al.*, 1946; reproduced with the permission of Williams & Wilkins

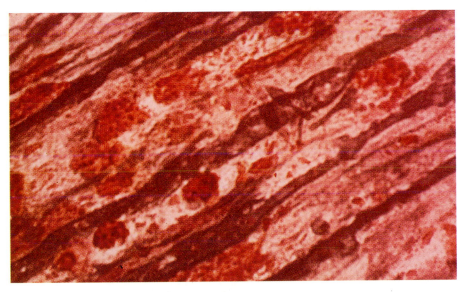

Figure 37 Histology (high-power view of the same section as in Figure 36) shows fat-containing macrophages (arrow) scattered among degenerating myelin sheaths. Combined myelin–fat stains. From Roizin *et al.*, 1946; reproduced with the permission of Williams & Wilkins

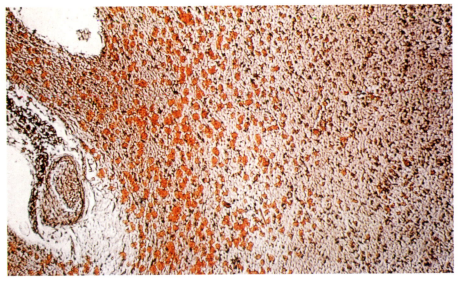

Figure 38 Histology (low-power view) of a fresh plaque from a patient with acute multiple sclerosis shows macrophages filled with 'neutral' fat. Oil red–O stain

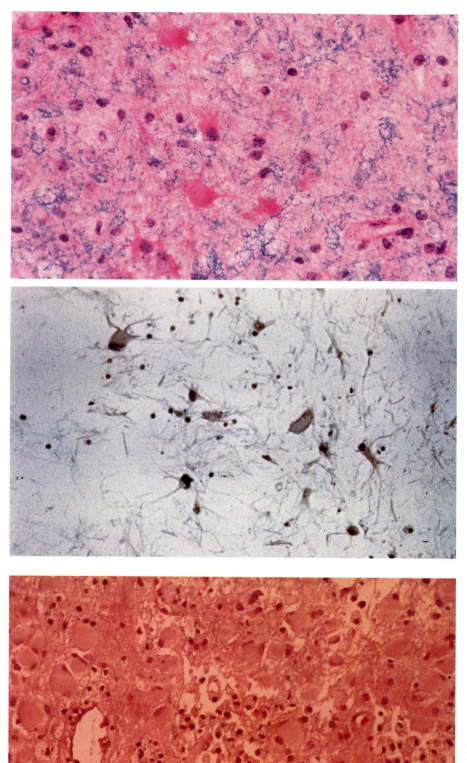

Figure 39 Histology of an early multiple sclerosis plaque shows gemistocytic astrocytes, seen with H & E–Luxol fast blue (upper) and Golgi (lower) staining

Figure 40 Biopsy specimen of a large cerebral plaque that was erroneously diagnosed as an astrocytoma. Note the large collection of macrophages. Periodic acid–Schiff (PAS) stain. (Figure 71 shows the CT scan of this patient)

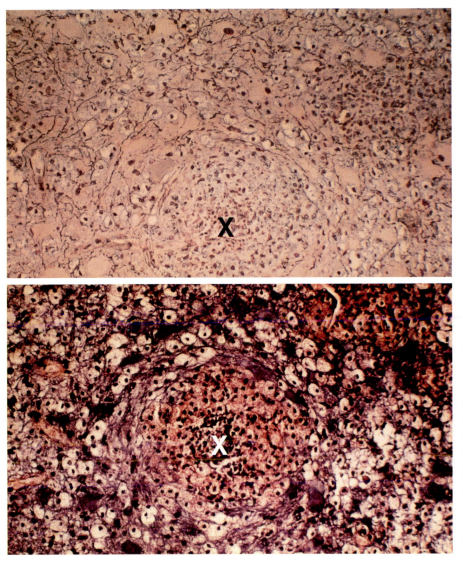

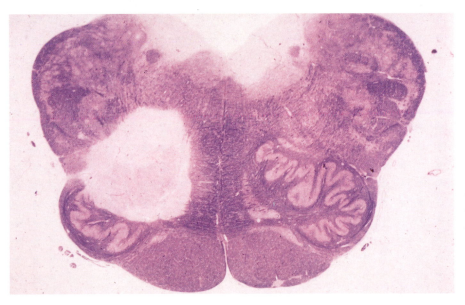

Figure 41 Biopsy specimen of a deep thalamic plaque misdiagnosed as a brain tumor, as seen by Bodian (upper) and phosphotungstic acid–hematoxylin (PTAH; lower) stains. There is an enormous accumulation of macrophages and lymphocytes in the perivascular space and surrounding tissue (**X** marks the blood vessel)

Figure 42 Section of the medulla shows a large lesion that has completely destroyed the left inferior olive. Luxol fast blue–PAS stain. Courtesy of Dr William Pendlebury, University of Vermont, Burlington, VT

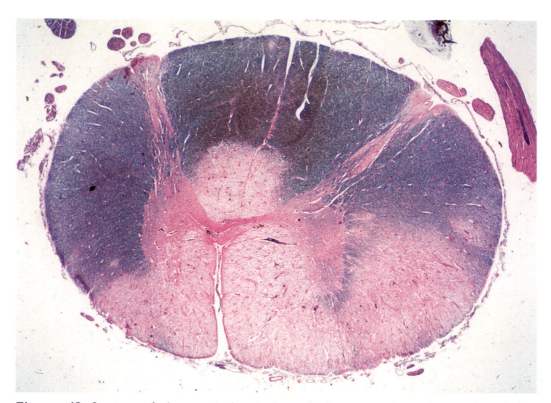

Figure 43 Section of the cervical spinal cord showing a large lesion. Luxol fast blue–PAS stain. Courtesy of Professor H. Lassmann, Institute of Brain Research, Vienna, Austria

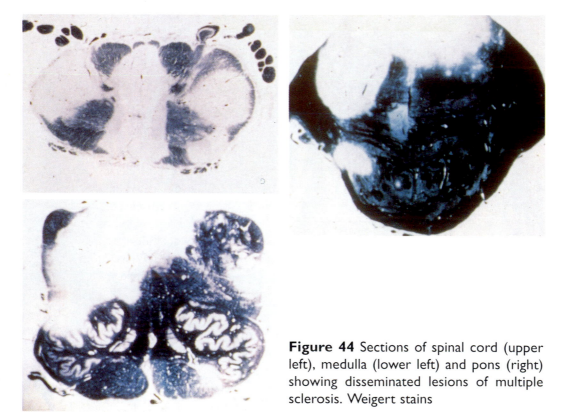

Figure 44 Sections of spinal cord (upper left), medulla (lower left) and pons (right) showing disseminated lesions of multiple sclerosis. Weigert stains

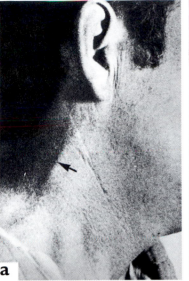

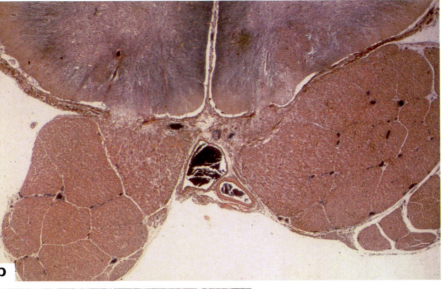

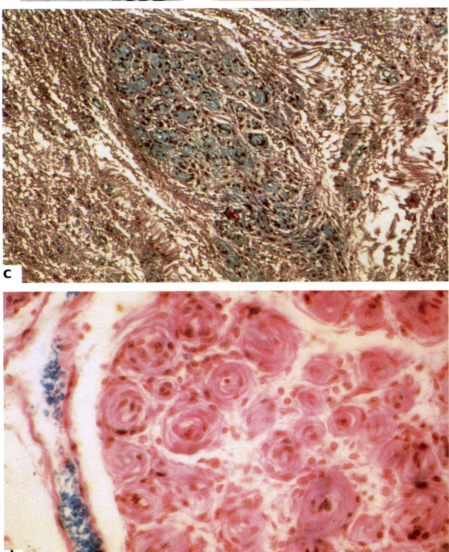

Figure 45 Dejerine–Sottas type of 'onion-bulb' enlargement of spinal roots and peripheral nerves in multiple sclerosis: (a) The greater auricular nerve is visibly enlarged; (b) section showing marked enlargement of the ventral roots; (c) Schwann cells of a peripheral nerve showing an onion-bulb appearance, seen by Gomori–trichrome stain; and (d) high-power view of onion-bulb Schwann cells (same section as in c), seen by H & E–Luxol fast blue stain. Courtesy of Thomas *et al.*, 1987, reproduced with the permission of Oxford University Press (a); and of Dr William Schoene, Brigham and Women's Hospital, Boston, MA (b–d)

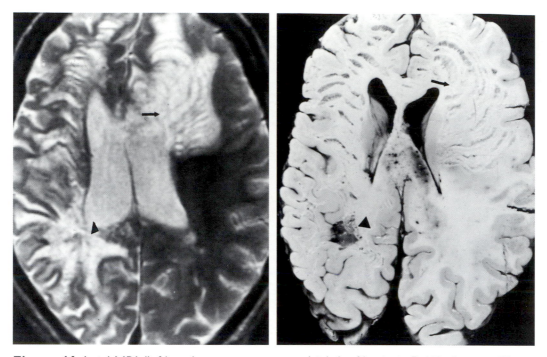

Figure 46 Axial MRI (left) and gross appearance (right) of brain in Baló's disease. There is concentric demyelination (arrows) and another lesion (indicated by the triangles). (Figure 85 b shows MRIs of the same patient.) Courtesy of Dr George Collins, State University of New York at Syracuse, from Gharagozloo *et al.*, 1994; reproduced with the permission of The Radiological Society of America

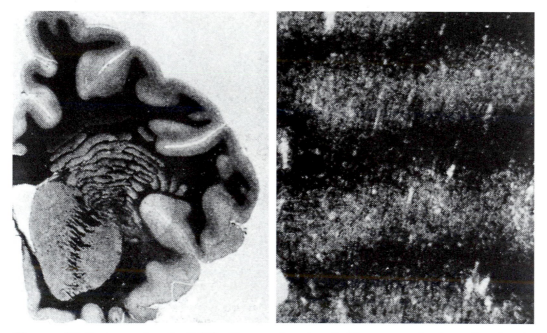

Figure 47 Brain section in Baló's disease showing (left) concentric demyelination in the right centrum semiovale. High-power view of the same area (right) shows alternating bands of normal myelin and demyelination. Weigert stains. Courtesy of Professor H. Shiraki, University of Tokyo, Japan

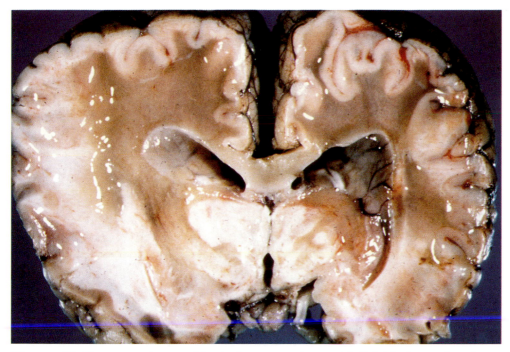

Figure 48 Gross specimen of brain from a patient with Schilder's myelinoclastic diffuse sclerosis shows large bilateral areas of demyelination

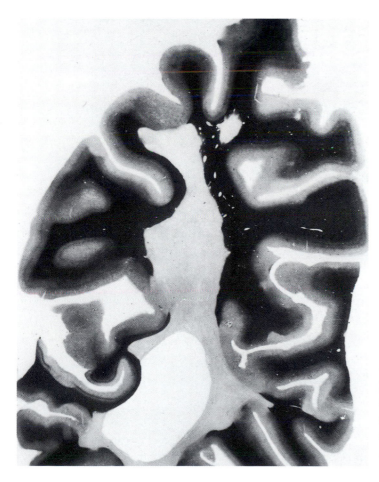

Figure 49 Celloidin section of brain from a patient with Schilder's myelinoclastic diffuse sclerosis shows a large plaque occupying most of the centrum semiovale. Weigert stain. Courtesy of Professor H. Shiraki, University of Tokyo, Japan

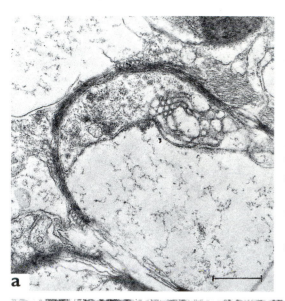

Figure 50 Electronmicrographs showing (a) vesicular dissolution of myelin in acute multiple sclerosis (bar = 0.5 μm); and, in chronic multiple sclerosis, (b) a naked axon (indicated by star) enveloped by a macrophage containing fragments of undegraded (native periodicity) myelin debris (arrows), indicative of recent uptake (bar = 1.0 μm); (c) various stages of myelin degradation in a macrophage (bar = 1.0 μm); (d) lyre bodies and lipid droplets in a macrophage (bar = 0.5 μm); (e) an isolated oligodendrocyte and myelin sheath among naked axons (indicated by stars) and astrocytic processes (arrows; bar = 2.0 μm); (f) periaxial segmental demyelination, in which loss of a myelin internodal segment has resulted in paranodal axolemmal specialization (between arrows). Filament-rich astrocytic processes surround the naked axon (bar = 2.0 μm); (g) subcortical white matter

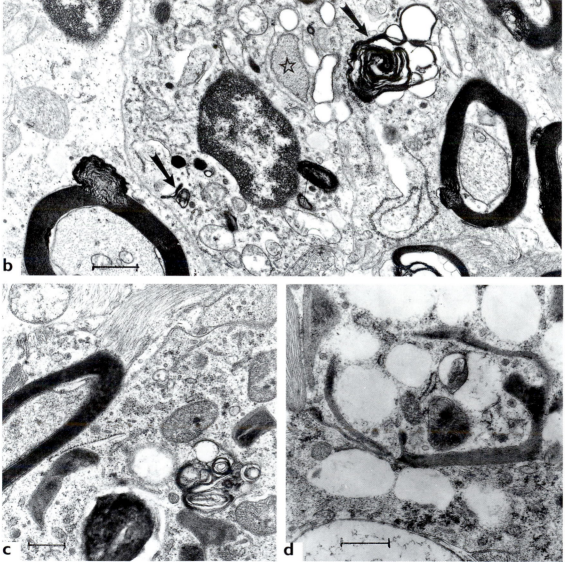

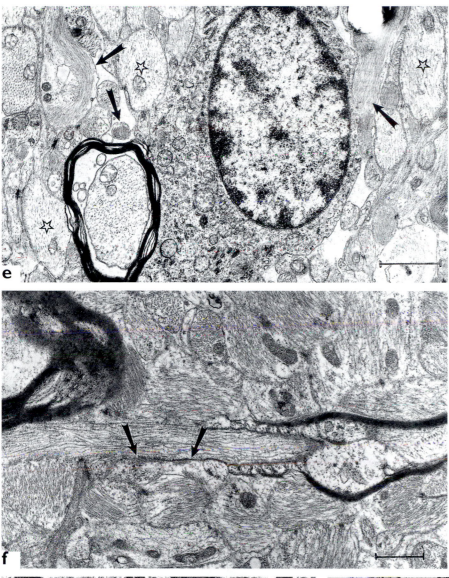

e

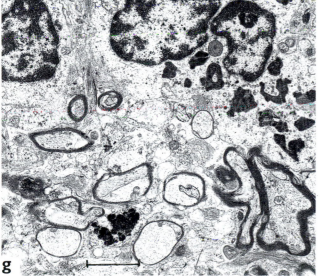

f

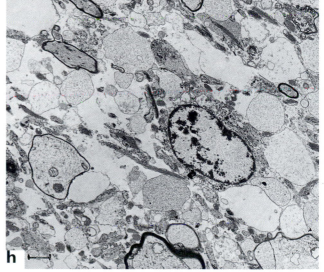

g

h

from an area of reduced myelin density at the edge of a plaque. Many of the axons show abnormally thin (?re-myelinated) myelin sheaths; macrophages are numerous, but do not contain recent (normal periodicity) myelin debris. Astrocytic gliosis is pronounced (bar = 2.0 μm); (h) a large periventricular plaque in which naked axons persist alongside thin remyel-inated axons and fine astro-cytic processes containing bundled filaments (bar = 2.0 μm).

Selected by Professor Ingrid Allen and Dr John Kirk, Queen's University of Belfast and The Royal Victoria Hospital, Belfast, Northern Ireland, UK. Courtesy of Dr John Kirk (a, b, e–h) and Dr Michael Hutchinson (c & d), with the permission of Black-well Science (a); from Allen, 1991, with the permission of Churchill Livingstone (b & e); and from Allen and Kirk, 1992, with the permission of Edward Arnold (c–e)

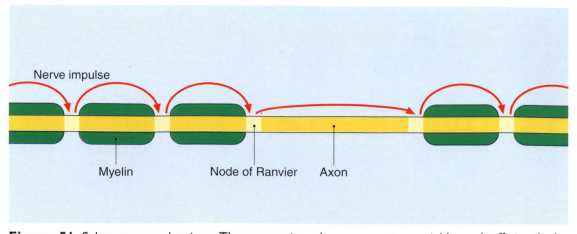

Figure 51 Saltatory conduction. The nerve impulse propagates quickly and efficiently by jumping from node to node. When segments of the myelin sheath are destroyed, the distance becomes too great and the impulse must then travel via the axon

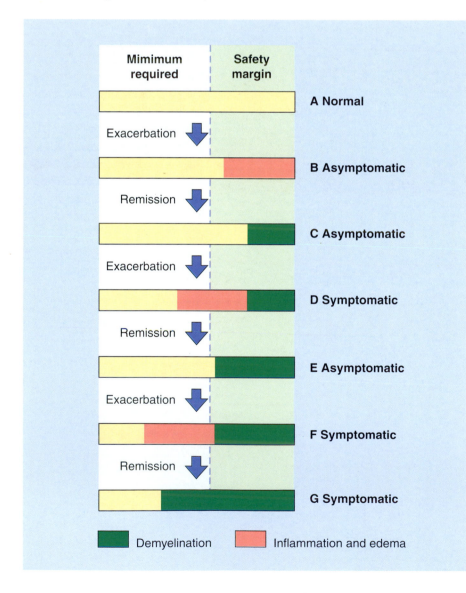

Figure 52 Safety factor in the progression of multiple sclerosis: As long as the required minimum number of nerve fibers remains intact, the patient is asymptomatic. When symptomatic myelin edema occurs during an exacerbation, remission may result in either complete or partial recovery, depending upon functional restoration of the required minimum number of fibers, or in a permanent deficit, as a result of myelinoclasia. From Poser, 1993; reproduced with the permission of Elsevier Science BV

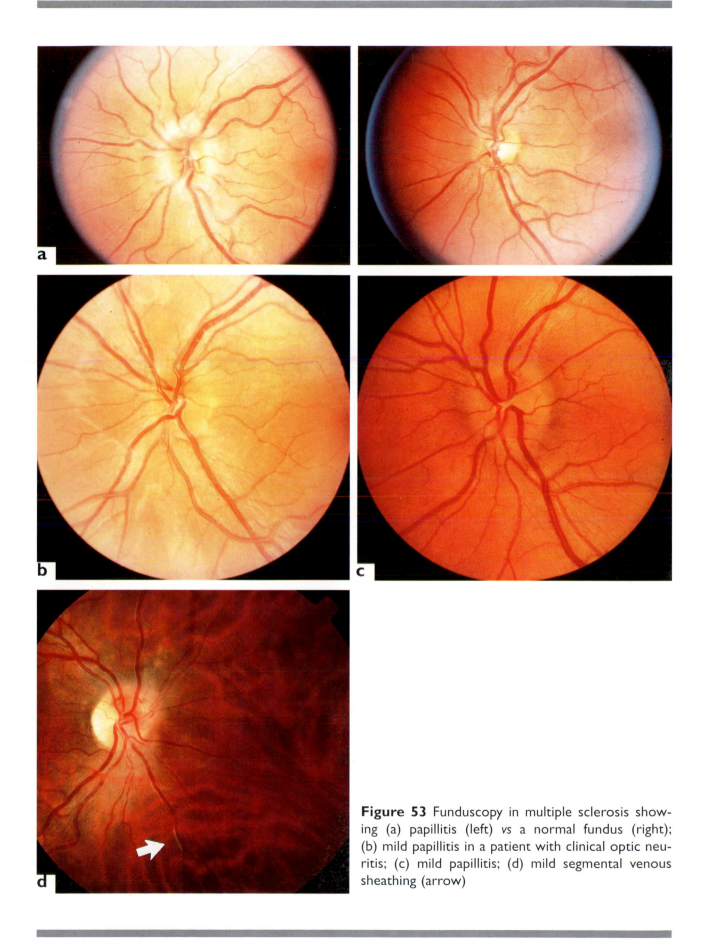

Figure 53 Funduscopy in multiple sclerosis showing (a) papillitis (left) *vs* a normal fundus (right); (b) mild papillitis in a patient with clinical optic neuritis; (c) mild papillitis; (d) mild segmental venous sheathing (arrow)

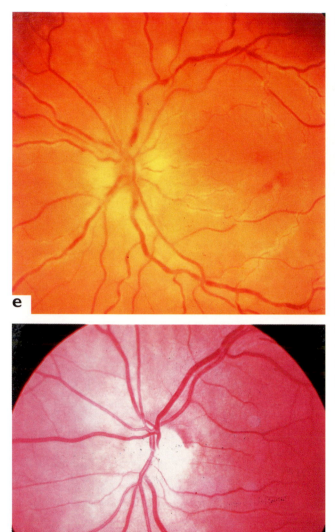

e

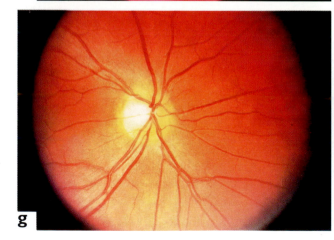

f

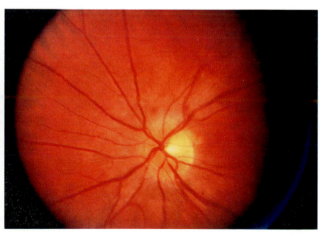

g

Figure 53 continued
(e) extensive periphlebitis; (f) temporal pallor after optic neuritis; (g) optic atrophy (left) *vs* a normal fundus (right); (h & i) severe optic atrophy; and (j) watercolor of a flattened retina showing the optic disk and macula (towards the center) as well as the most anterior portion of the retina, the pars plana (the scalloped yellow margin), which lies just behind the ciliary body. Periphlebitis is evident in some of the distal portions of the retinal veins. Inflammatory cells in the vitreous have settled over the inferior retina, giving the appearance of 'snowbanks'. Severe cases may also have cystoid macular edema, as shown here.

Courtesy of Dr Jason Barton, Beth Israel Deaconess Medical Center, Boston, MA (d); Dr James Folk, University of Iowa, Iowa City (e); and Drs Janet Davis, Bascom–Palmer Eye Institute, and Noble David, University of Miami, FL (j); reproduced with the permission of the American Academy of Ophthalmology. All other figures are courtesy of Dr Simmons Lessell, Massachusetts Eye and Ear Infirmary, Boston, MA

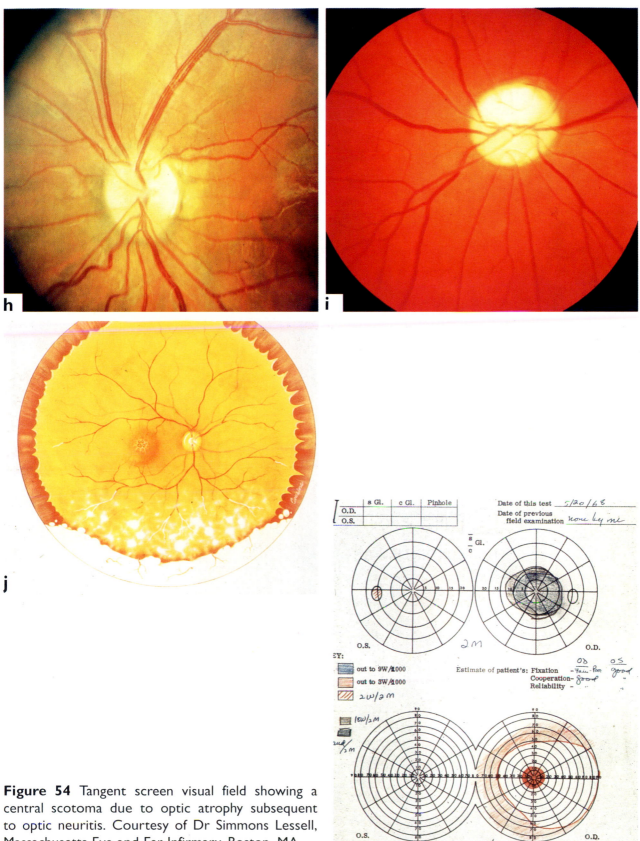

Figure 54 Tangent screen visual field showing a central scotoma due to optic atrophy subsequent to optic neuritis. Courtesy of Dr Simmons Lessell, Massachusetts Eye and Ear Infirmary, Boston, MA

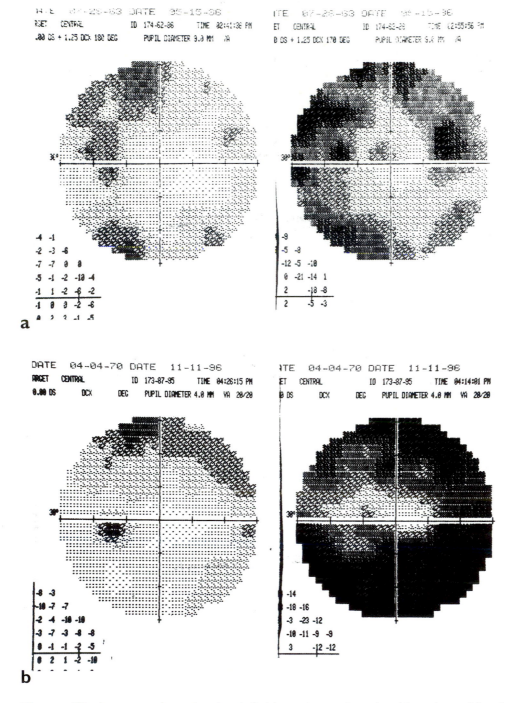

Figure 55 Automated static visual field testing using the Humphrey Visual Field Analyzer shows: (a) Central scotoma of the right eye in optic neuritis; (b) central field sparing of the right eye with severe peripheral constriction in optic neuritis

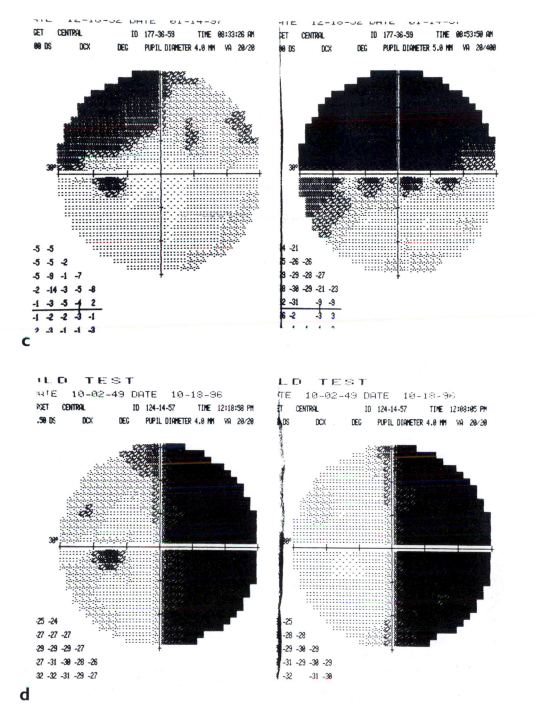

c

d

Figure 55 continued
(c) inferior chiasmal lesion; and (d) lesion of the left optic tract. Courtesy of
Dr Thomas Hedges, Tufts–New England Medical Center, Boston, MA. For technical details, see Keltner *et al.*, 1993

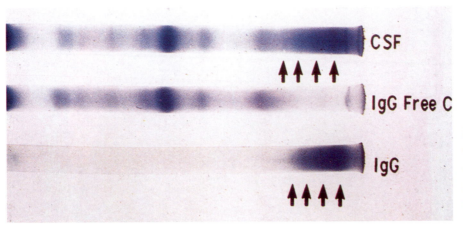

Figure 56 Electrophoresis of cerebrospinal fluid showing increased levels of immuno-globulin G (arrows), a non-specific finding

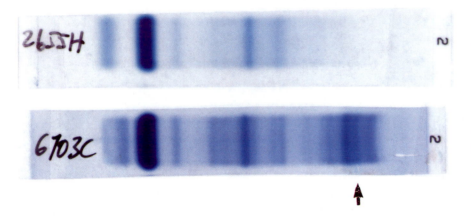

Figure 57 Electrophoresis of cerebrospinal fluid showing (upper) normal *vs* (lower) ×80 increased concentration; three oligoclonal bands can be seen in the IgG fraction (arrow). Coomassie brilliant blue stain. Courtesy of Dr James Faix, Beth Israel Deaconess Medical Center, Boston, MA

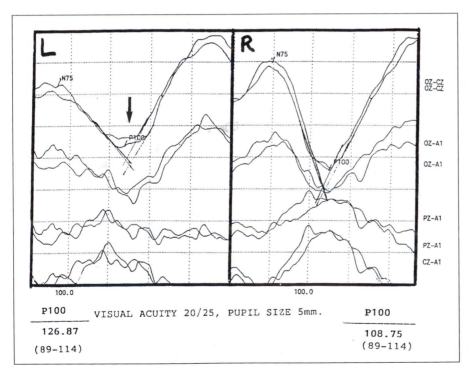

Figure 58 Visual evoked response in a patient with left optic atrophy. The P100 wave on the left is delayed compared with that on the right. Courtesy of Dr Frank Drislane, Beth Israel Deaconess Medical Center, Boston, MA

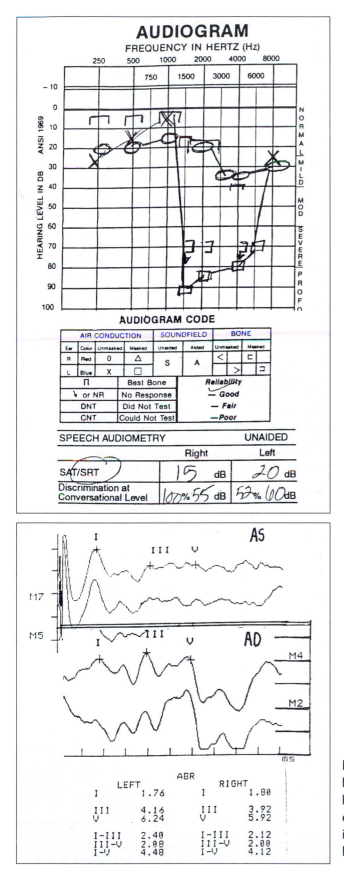

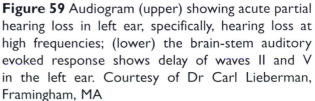

Figure 59 Audiogram (upper) showing acute partial hearing loss in left ear, specifically, hearing loss at high frequencies; (lower) the brain-stem auditory evoked response shows delay of waves II and V in the left ear. Courtesy of Dr Carl Lieberman, Framingham, MA

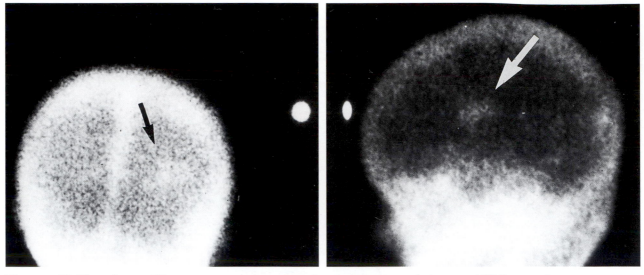

Figure 60 Phosphorus-32 scans, examples of early neuroimaging (taken in 1954) reveal a large plaque in the posterior frontal subcortical white matter (arrows). Courtesy of Dr Edward Schlesinger, The New York Neurological Institute, NY

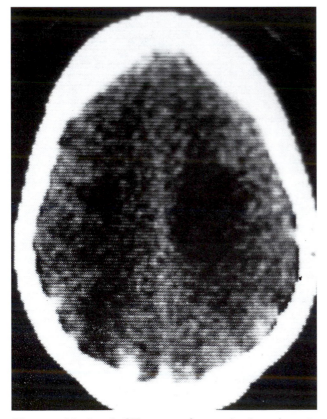

Figure 61 Axial CT scan of a multiple sclerosis patient with an acute left hemiparesis showing a large area of hypodensity in the right hemisphere, and two smaller areas in the left

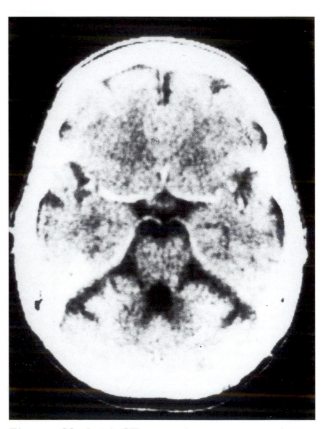

Figure 62 Axial CT scan showing severe brainstem and cerebellar atrophy

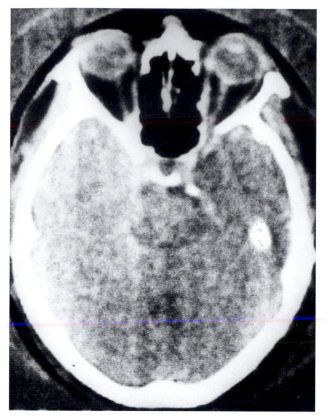

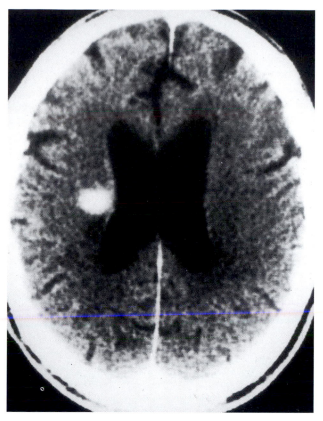

Figure 63 Axial CT scan showing bilateral optic atrophy

Figure 64 Axial CT scan, using iodinated contrast, shows a single periventricular plaque on the left. The diagnosis was confirmed by post-mortem examination

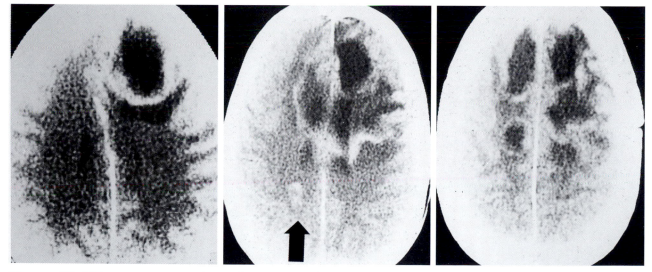

Figure 65 CT scans, using iodinated contrast, show the development of malignant multiple sclerosis over time: (left) Initially, there is a large left frontal lesion with ring enhancement suggestive of abscess; (middle) 4 months later, there is bilateral extension and new plaque posteriorly (arrow); (right) after 2 more weeks, there is further development of plaques bilaterally. The diagnosis was confirmed by post-mortem examination

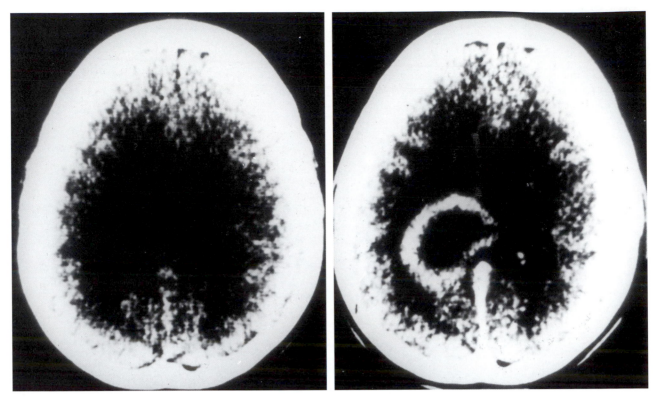

Figure 66 Axial CT scans, using iodinated contrast, taken after 30 min (left) and after a delay of 60 min (right). After 30 min, no lesions can be seen whereas, after the 1-hour delay, there is ring enhancement suggestive of brain abscess. The diagnosis of multiple sclerosis was confirmed by biopsy. (As these images were taken in 1978, the resolution is poor compared with the results using more recent technology.)

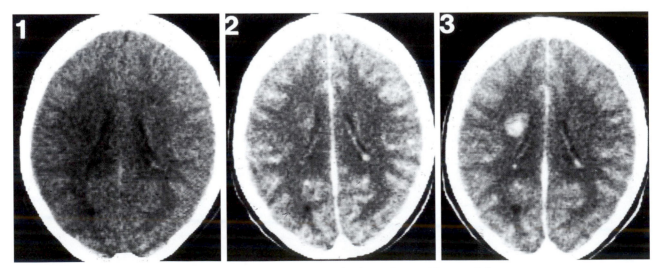

Figure 67 Axial CT scans, using iodinated contrast and taken after a 60-min delay, compare the effects of (1) no contrast with (2) 300 mg and (3) 600 mg of contrast. With 300 mg, the left posterior frontal lesion is only faintly seen whereas doubling the dose renders the lesion much more obvious. Courtesy of Drs Fernando Vinuela and George Ebers, University of Western Ontario, London, Canada

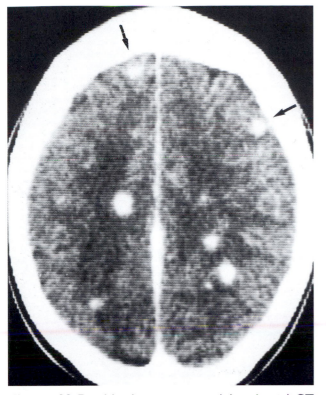

Figure 68 Double-dose contrast delayed axial CT scan shows multiple lesions, including two in the cortex (arrows)

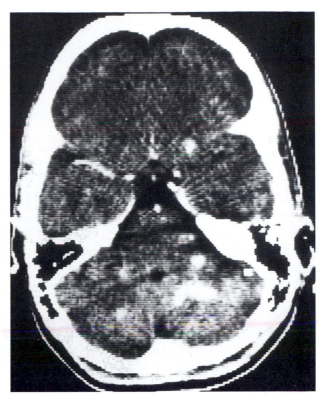

Figure 69 Double-dose contrast delayed axial CT scan shows multiple lesions in the cerebellum

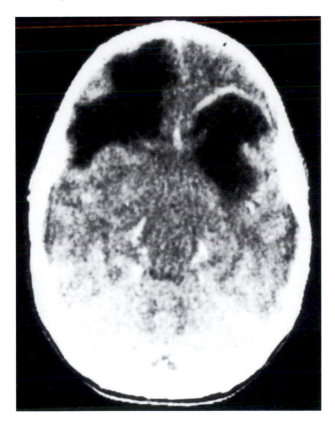

Figure 70 Axial CT scan, using iodinated contrast, of a patient with Schilder's myelinoclastic diffuse sclerosis shows large bilateral frontal lesions with ring enhancement

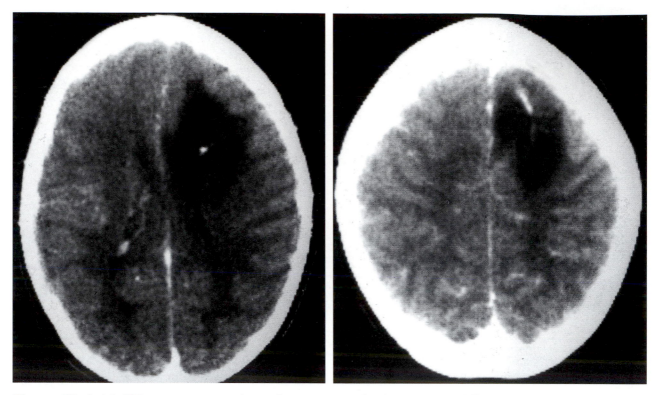

Figure 71 Axial CT scans, using iodinated contrast and taken at two different planes, show a large right frontal lesion. This had been erroneously diagnosed as astrocytoma on biopsy (see Figure 40)

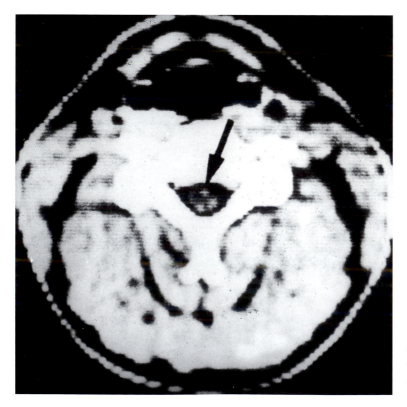

Figure 72 Axial CT scan, using iodinated contrast, shows lesions (arrow) at the level of C4 of the spinal cord. Plaques of the spinal cord are rarely seen by CT

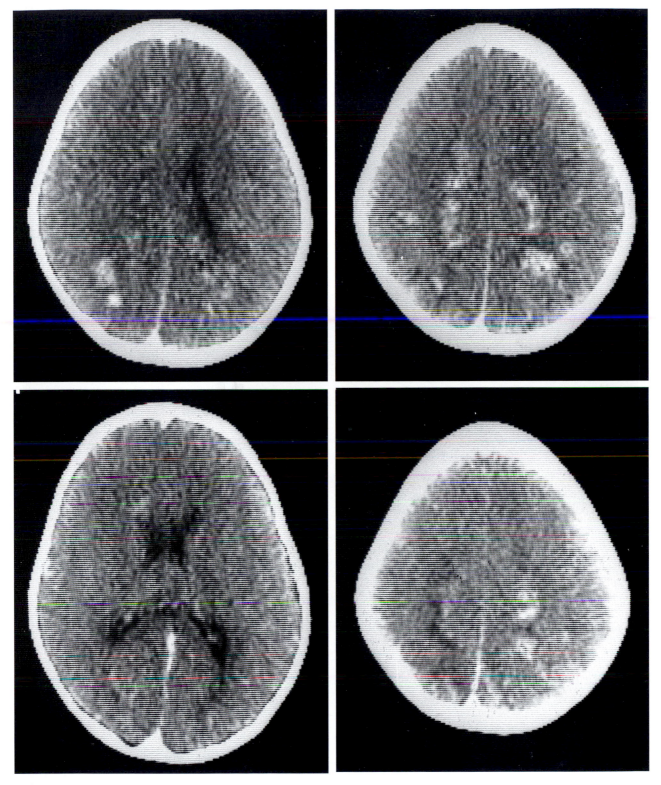

Figure 73 CT scans, using iodinated contrast, of a patient who had acute disseminated encephalomyelitis (ADEM). This had been misdiagnosed, on the basis of clinical history, as multiple sclerosis

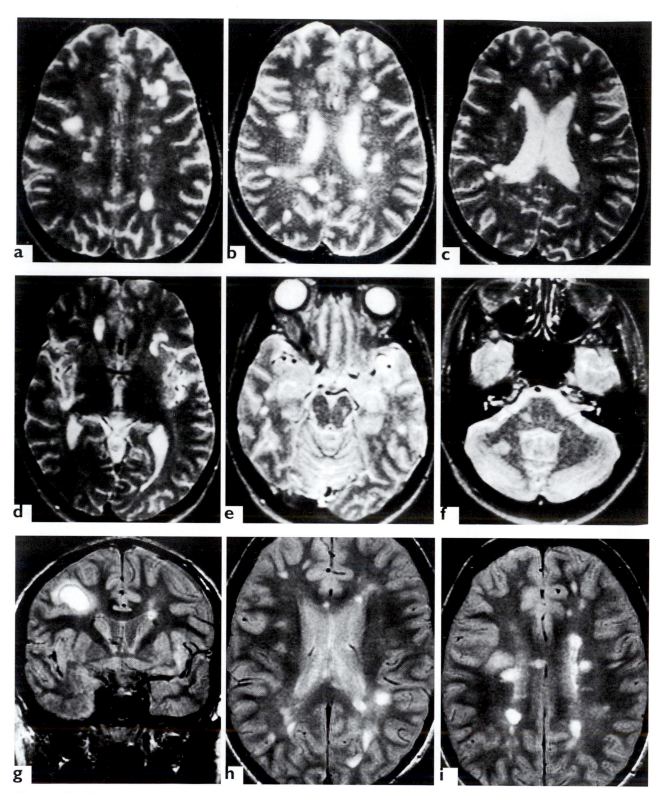

Figure 74 T$_2$-weighted (a–j) and proton-density (k–r) MRIs showing areas of increased signal intensity (AISI); these are frequently seen in clinically definite multiple sclerosis, but are not diagnostic. Note the apparent recent enlargement of the dense right subcortical AISI in **j** (arrow)

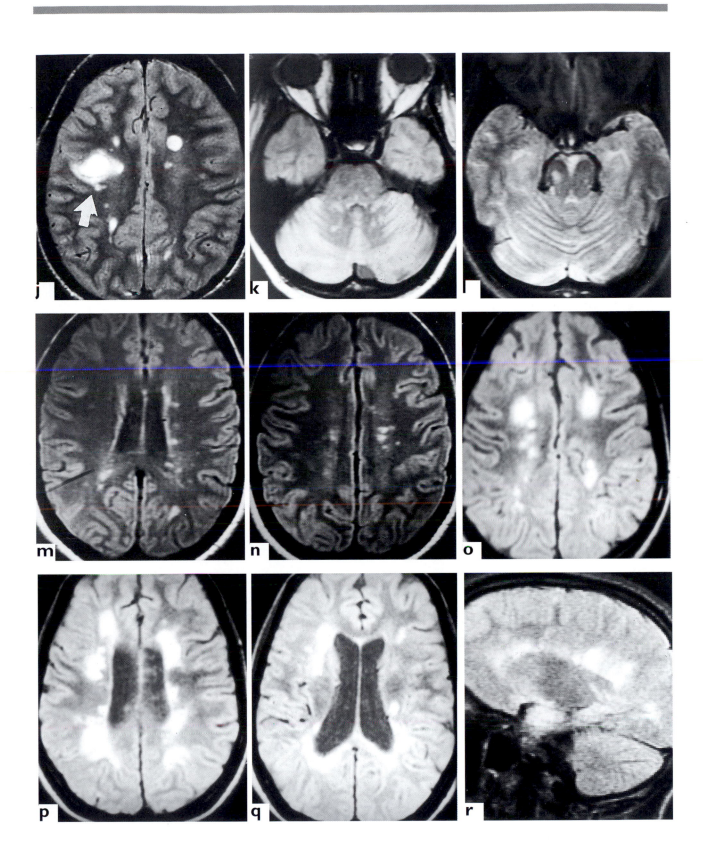

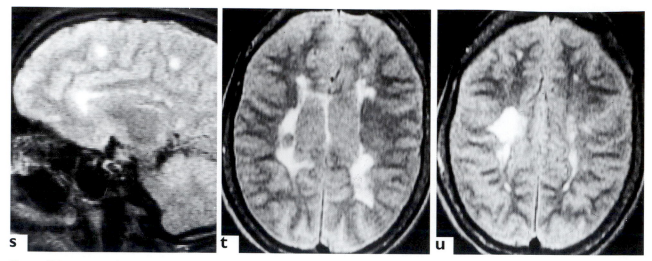

Figure 74 continued
Proton-density MRIs (s–u) showing areas of increased signal intensity (AISI); these are frequently seen in clinically definite multiple sclerosis, but are not diagnostic

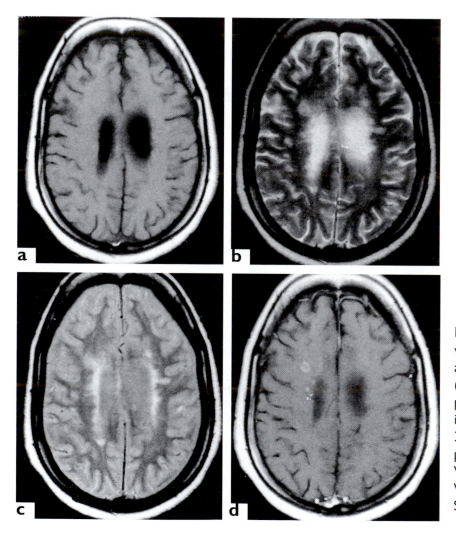

Figure 75 T_1-weighted (a), T_2-weighted (b), proton-density (c) and gadolinium–EDTA-enhanced (d) MRIs of the first histologically proven case of multiple sclerosis in a black South African. (Figure 30 shows the brain biopsy of this patient.) Courtesy of Professor Vivian Fritz, University of the Witwatersrand, Johannesburg, South Africa

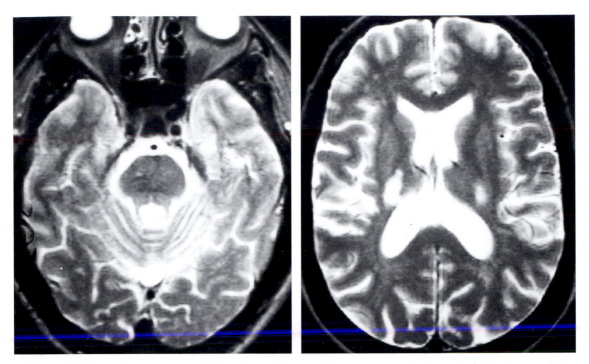

Figure 76 T$_2$-weighted axial MRIs of a 65-year-old woman with clinically definite multiple sclerosis. Note the bilateral, almost symmetrical, areas of increased signal intensity

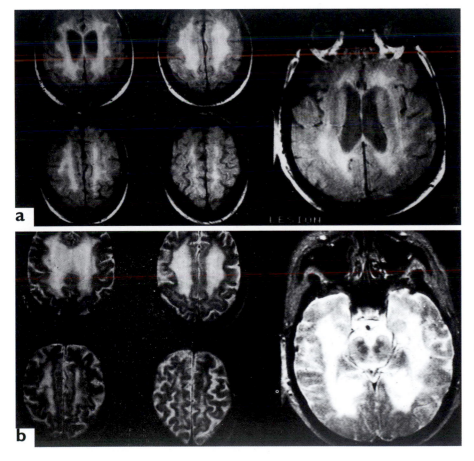

Figure 77 T$_2$-weighted (a) and proton-density (b) axial MRIs of a patient with multiple sclerosis and severe dementia. There is nearly total demyelination bilaterally in the centrum semiovale

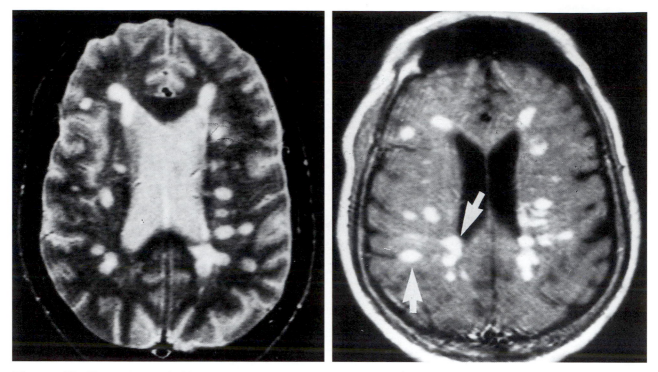

Figure 78 T$_2$-weighted (left) and gadolinium–EDTA-enhanced (right) axial MRIs showing a number of ovoid lesions. Although many areas of increased signal intensity are seen on T$_2$-weighting, other such areas (arrows) not revealed by T$_2$-weighting are clearly seen with gadolinium–EDTA enhancement. This phenomenon is relatively uncommon

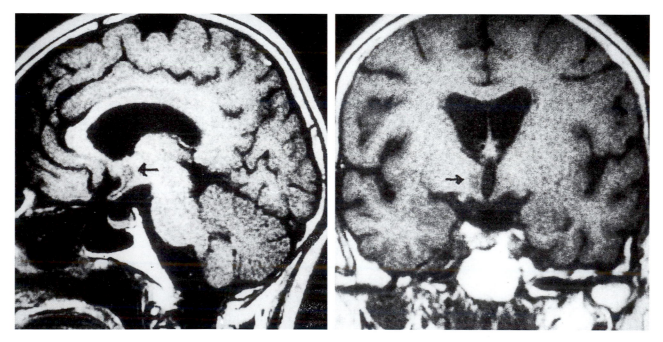

Figure 79 T$_1$-weighted sagittal (left) and coronal (right) MRIs of a multiple sclerosis patient with prolactinemia. Note the lesion in the hypothalamus and pituitary stalk (arrows)

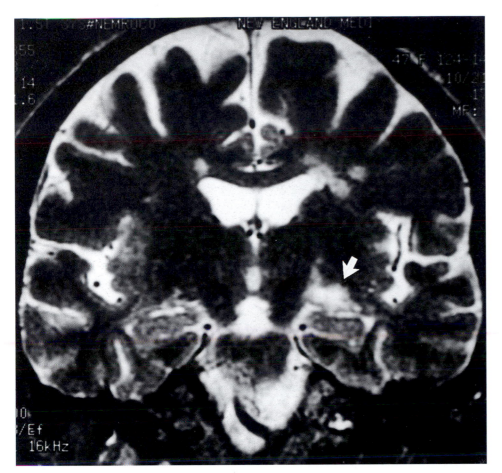

Figure 80 T$_2$-weighted coronal MRI showing areas of increased signal intensity in the left optic tract (arrow) and at the superior angles of the ventricles

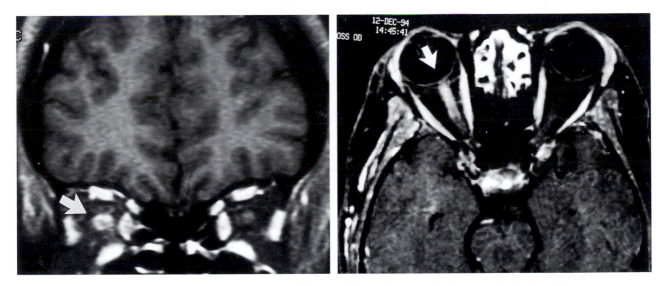

Figure 81 T$_2$-weighted coronal (left) and axial (right) MRIs showing lesions resulting in right optic neuritis (arrows). Courtesy of Drs Simmons Lessell and Judith Warner, Massachusetts Eye and Ear Infirmary, Boston, MA

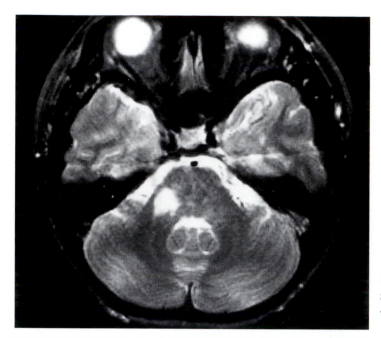

Figure 82 MRI of the brain stem shows areas of increased signal intensity in a patient with right facial palsy

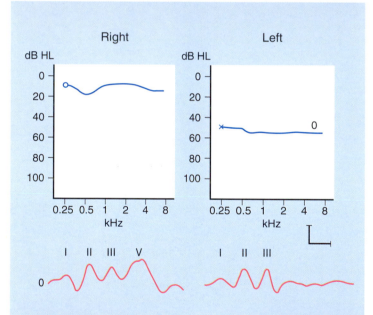

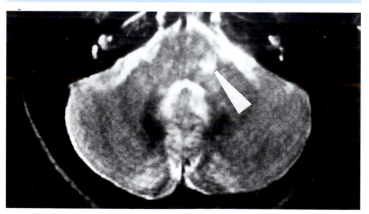

Figure 83 Audiogram (upper), auditory evoked response (middle) and MRI (lower) of a patient who had a sudden partial loss of hearing in the left ear. The audiogram shows decreased perception of all frequencies; the brain-stem auditory evoked response shows absence of wave V; and the MRI shows areas of increased signal intensity in the left brain stem just above the superior olive (arrow)

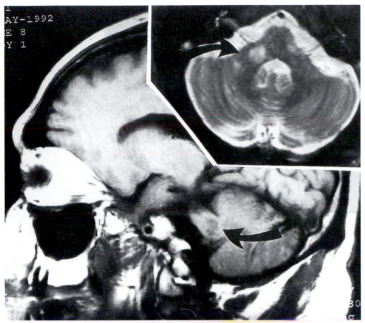

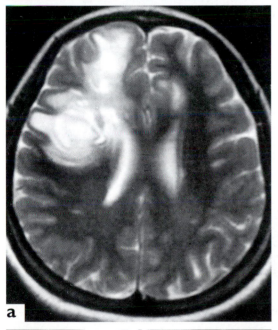

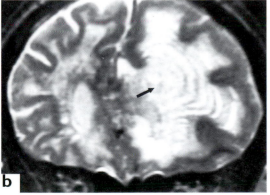

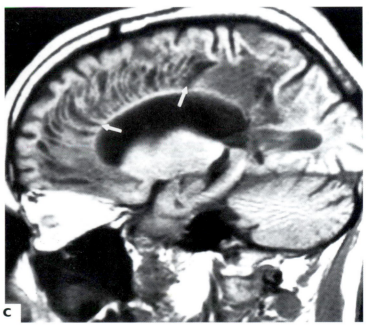

Figure 84 T$_1$-weighted MRIs of the cerebellum show areas of increased signal intensity (arrows) in the middle cerebellar peduncle

Figure 85 MRI appearances of Baló's disease. Courtesy of Dr Chi-Jen Chen, Chang-Gung Medical College, Taipei, Taiwan, from Chen *et al.*, 1996, reproduced with the permission of Lippincott–Raven (a); and Dr George Collins, State University of New York at Syracuse, from Gharagozloo *et al.*, 1994, reproduced with the permission of The Radiological Society of America (b & c). (Figure 46 shows the pathology of this case)

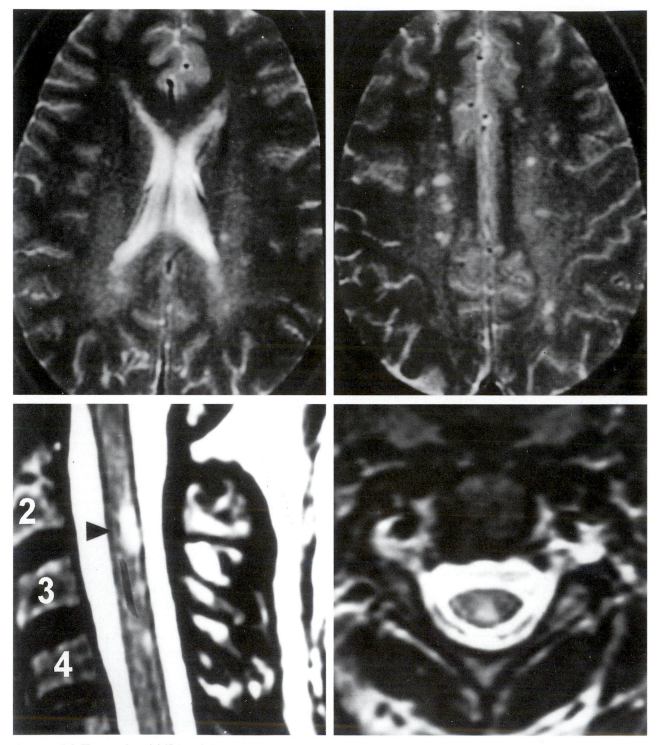

Figure 86 T$_2$-weighted MRIs of the brain (upper) and spinal cord (lower) of a patient who had no neurological signs or symptoms, but was suspected of having multiple sclerosis on the basis of the MRI and the presence of three oligoclonal bands in the cerebrospinal fluid. Note the areas of increased signal intensity at the level of the C2–3 intervertebral space (arrow), and the two smaller ones at the level of C3 and C4

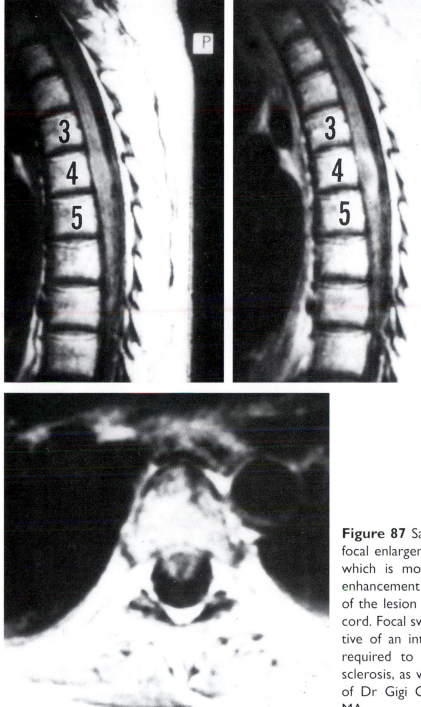

Figure 87 Sagittal MRIs of the spinal cord show focal enlargement at the level of C4 (upper left), which is more obvious with gadolinium–EDTA enhancement (upper right). The axial view (lower) of the lesion shows that the plaque encircles the cord. Focal swelling of the cord is strongly suggestive of an intramedullary tumor. Biopsy may be required to confirm the diagnosis of multiple sclerosis, as was necessary in this case. Courtesy of Dr Gigi Girgis, Waltham Hospital, Waltham, MA

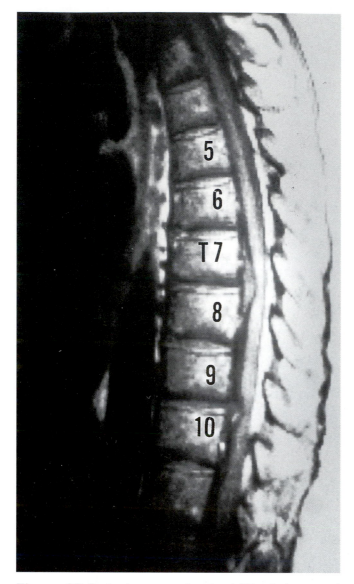

Figure 88 Sagittal proton-density MRI of the spinal cord shows a completely extruded disk compressing the cord at the level of T7–8. Note the areas of increased signal intensity extending both above and below the site of compression. There is also cord compression at the T9–10 level. The patient had suffered from a fall around 2 months previously

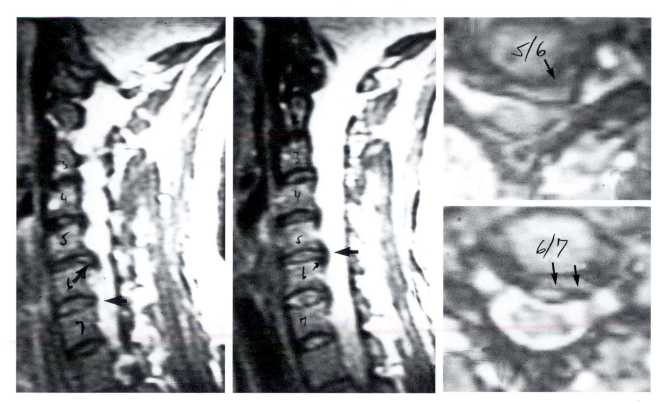

Figure 89 Sagittal (left and middle) and axial (right upper and lower) T₂-weighted MRIs show herniated disks and spondylosis at multiple levels. The lesions are most severe at C5–6 and C6–7 (arrows; left and middle). The areas of increased signal intensity and the extruded disks are best seen in the axial views (arrows)

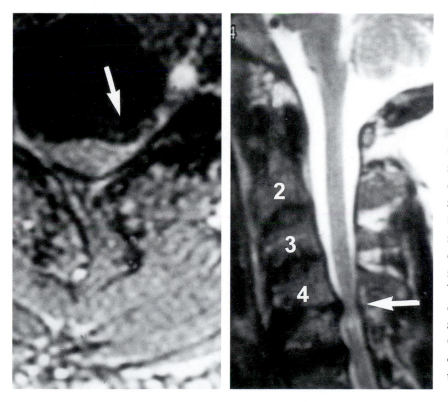

Figure 90 Axial (left) and sagittal (right) T₂-weighted MRIs show severe spondylosis at C4–5 with cord compression. Note the areas of increased signal intensity immediately below the compression site (arrow; right). Spondylosis at C5–6 has obliterated the subarachnoid space anteriorly. The cord is shoved backwards, causing narrowing of the subarachnoid space posteriorly from C3–4 downwards. The axial view of C4–5 shows compression of the cord clearly

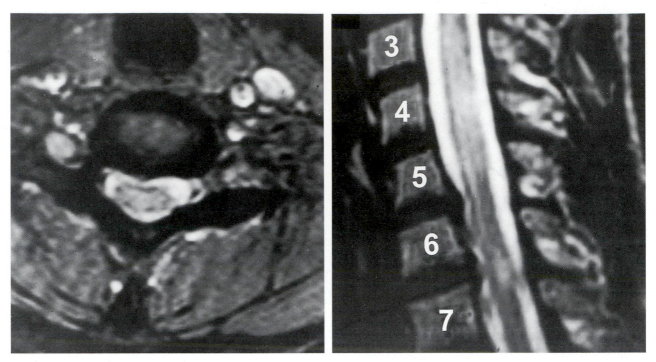

Figure 91 Axial (left) and sagittal (right) T$_2$-weighted MRIs show severe attenuation of the anterior subarachnoid space at C5–6 and C6–7. There is an area of increased signal intensity at the C6–7 level and evidence of distortion of the cord in the axial view. There is a close and almost certain causal relationship between the cord compression, and the multiple sclerosis plaques in this patient and in those depicted in Figures 88–90

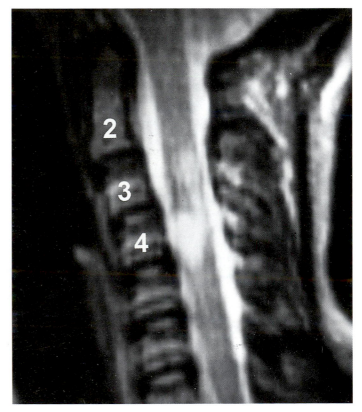

Figure 92 Sagittal T$_2$-weighted MRI taken 8 months after a whiplash injury to the neck shows loss of the normal cervical lordosis due to cervical muscle spasm. A posterior area of increased signal intensity is seen at the level of C3; another, larger and more dense, lies anterior to and below it. There is attenuation of the anterior subarachnoid space at C4–5, C5–6 and C6–7 levels

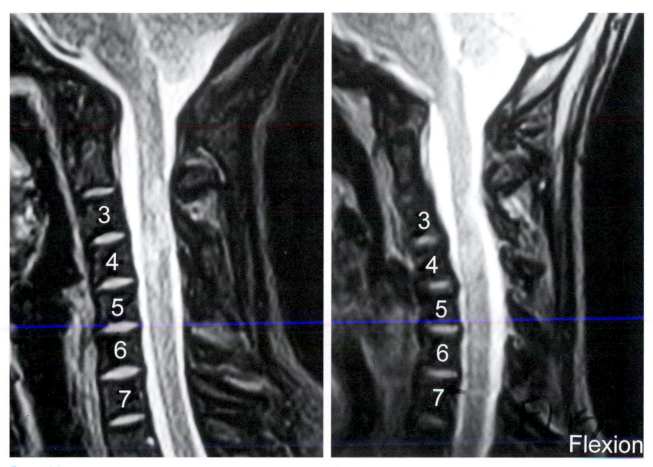

Figure 92 continued
Sagittal T$_2$-weighted MRIs obtained 14 months later show less muscle spasm, and the area of increased signal intensity at C4 appears to be less dense. There is attenuation of the subarachnoid space posteriorly at the level of C3 due to a fold in the ligament, which corresponds to the area of increased signal intensity at that level.

With the neck in flexion (right), the anterior subarachnoid space is markedly attenuated, especially at C4–5 and C5–6 levels, whereas the posterior space is greatly enlarged presumably as a result of pressure on the cerebrospinal fluid surrounding the tethered cord. The posterior indentation at C3 persists. It is thought that violent hyperflexion of the neck at the time of the whiplash injury was responsible for the formation of the C4 plaque. Courtesy of Dr Gerald O'Reilly, CHEM MRI, Stoneham, MA

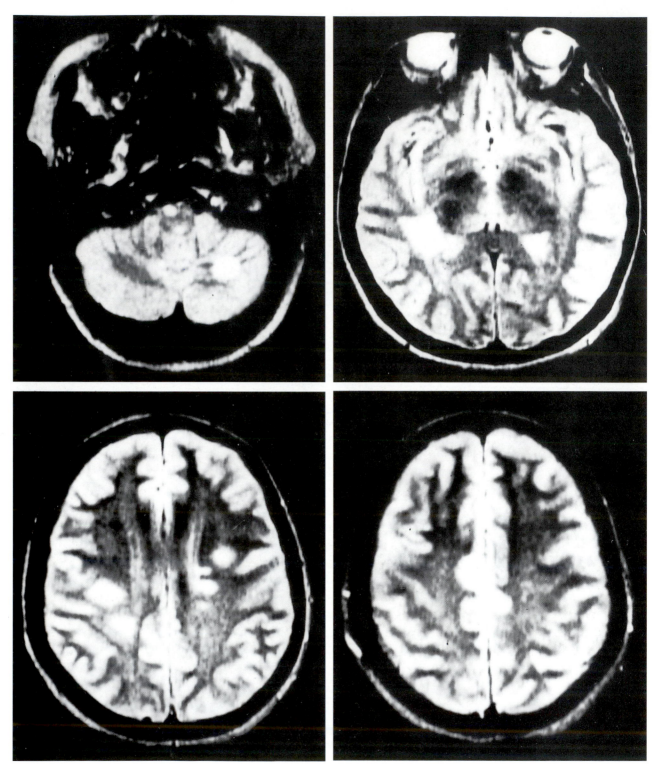

Figure 93 T_2-weighted axial MRIs of a patient with acute disseminated encephalomyelitis (ADEM) appear to fulfill the Fazekas *et al.,* 1988 criteria for multiple sclerosis. At least four areas of increased signal intensity $> 5\,$mm are present, located in the periventricular white matter and in the posterior fossa. From Poser, 1994a; reproduced with the permission of Lippincott–Raven

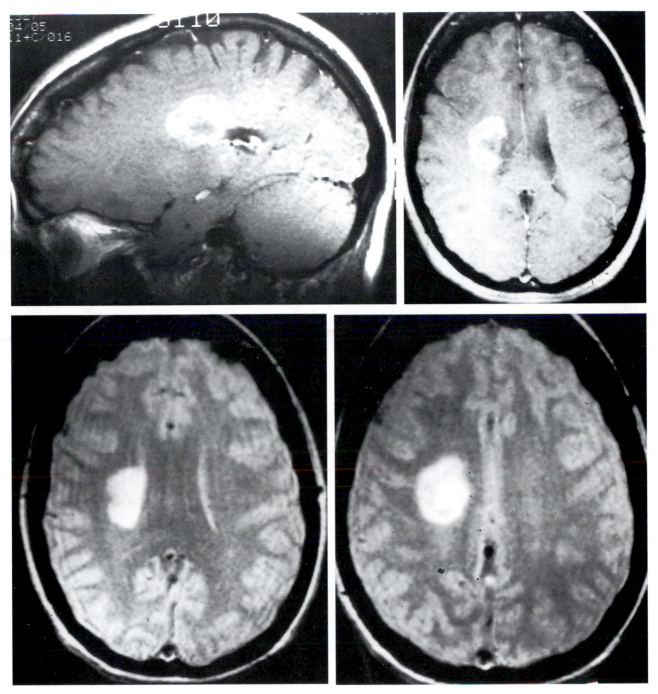

Figure 94 Sagittal (upper left) and axial (upper right and lower) proton-density MRIs of a patient with ADEM showing a single right periventricular area of increased signal intensity

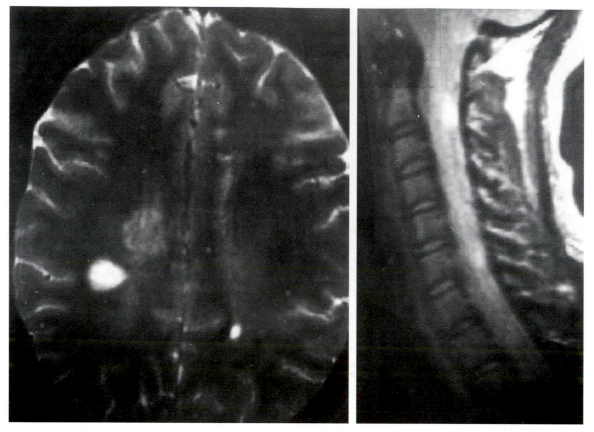

Figure 95 Proton-density MRIs of the brain (left; axial) and spinal cord (right; sagittal) of a patient with ADEM. The brain scan shows a dense area of increased signal intensity (AISI) in the right posterior parietal lobe and, lying medial and anterior to it, another AISI that is relatively faint. A small dense AISI is present in the left lobe. In the cervical cord, at least two AISI can be seen. The patient had been misdiagnosed as multiple sclerosis

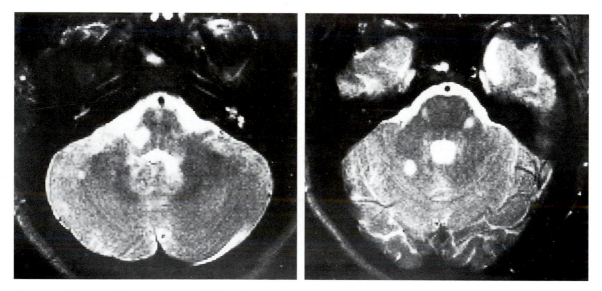

Figure 96 Axial T$_2$-weighted MRIs of a patient with ADEM showing areas of increased signal intensity restricted to the cerebellum

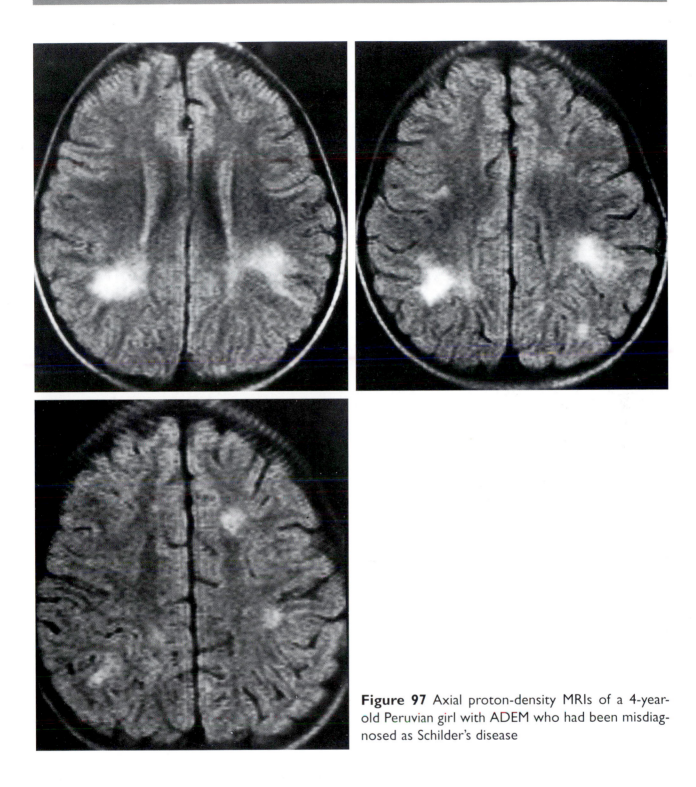

Figure 97 Axial proton-density MRIs of a 4-year-old Peruvian girl with ADEM who had been misdiagnosed as Schilder's disease

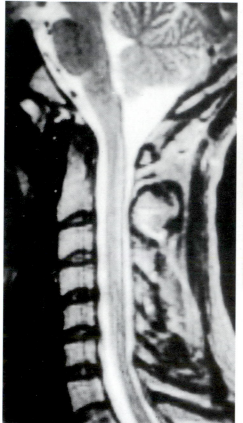

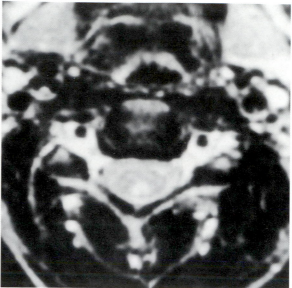

Figure 98 Sagittal (left) and axial (right) T$_2$-weighted MRIs of a patient with ADEM showing a single, very long, area of increased signal intensity restricted to the posterior columns of the upper cervical cord. The patient's only complaint was clumsiness due to loss of positional sense in her hands

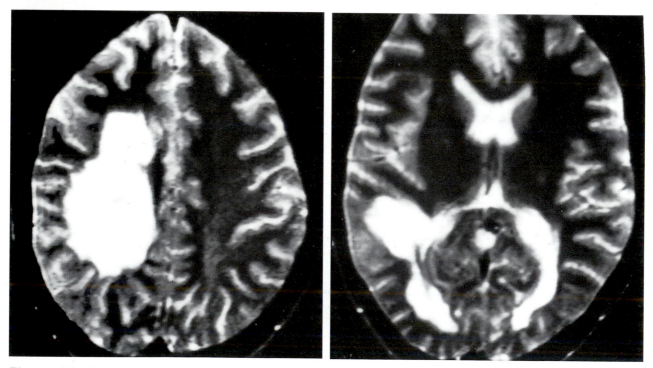

Figure 99 Coronal T$_2$-weighted MRIs of a patient with postvaccination (influenza) encephalomyelitis

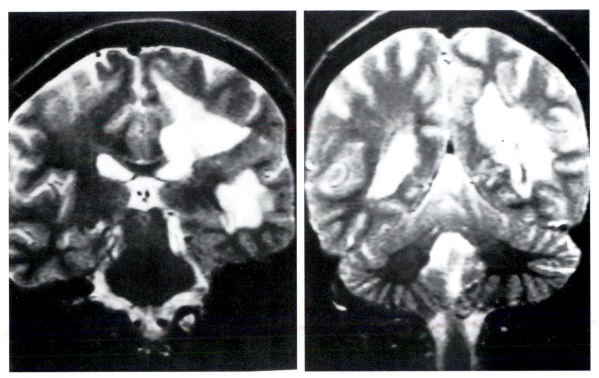

Figure 99 continued
Axial T$_2$-weighted MRIs of a patient with postvaccination (influenza) encephalomyelitis

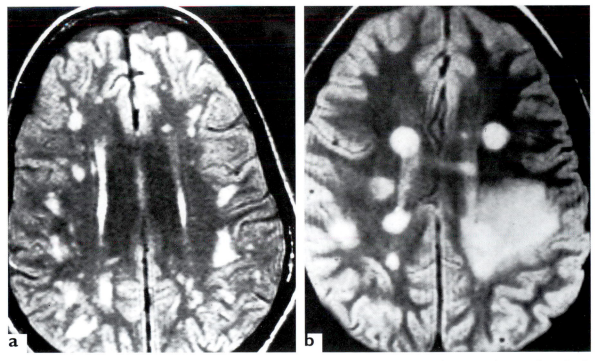

Figure 100 MRIs of eight patients mistakenly diagnosed as multiple sclerosis who, in fact, had: (a) ADEM, with areas of increased signal intensity (AISI) in the immediate subcortical white matter; (b) ADEM

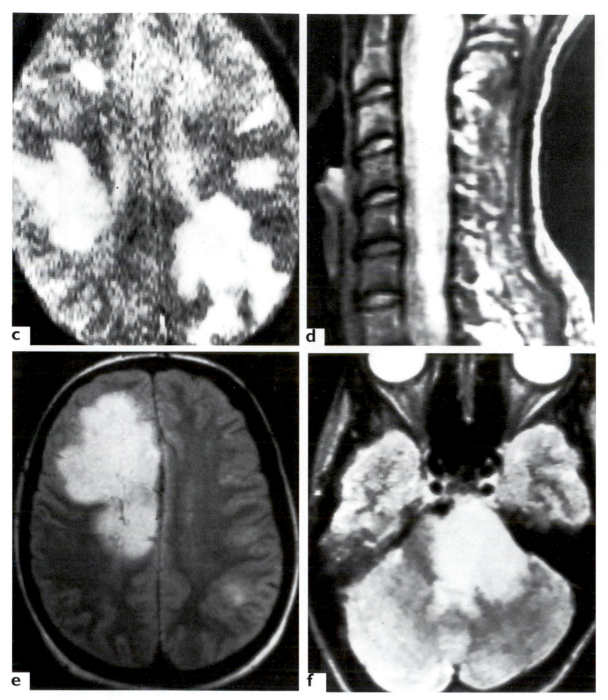

Figure 100 continued
(c) ADEM, with a large bilateral AISI involving the cortex (see also Figure 101); (d) ADEM, with acute transverse myelitis; (e) postvaccination (influenza) ADEM. Such large lobar AISI are never seen in multiple sclerosis; (f) ADEM. Such large AISI, seen here in the pons and cerebellum, are never seen in multiple sclerosis; (g) ADEM, with a single globular AISI in the posterior angle of the ventricle. A single AISI of this type is not uncommonly seen in ADEM; (h) ADEM in a patient with postvaccination (hepatitis B) lupus erythematosus who had sudden-onset aphasia following an upper respiratory infection several months later. A tumor was suspected, but biopsy showed demyelination. From Poser, 1994a; reproduced with the permission of Lippincott–Raven

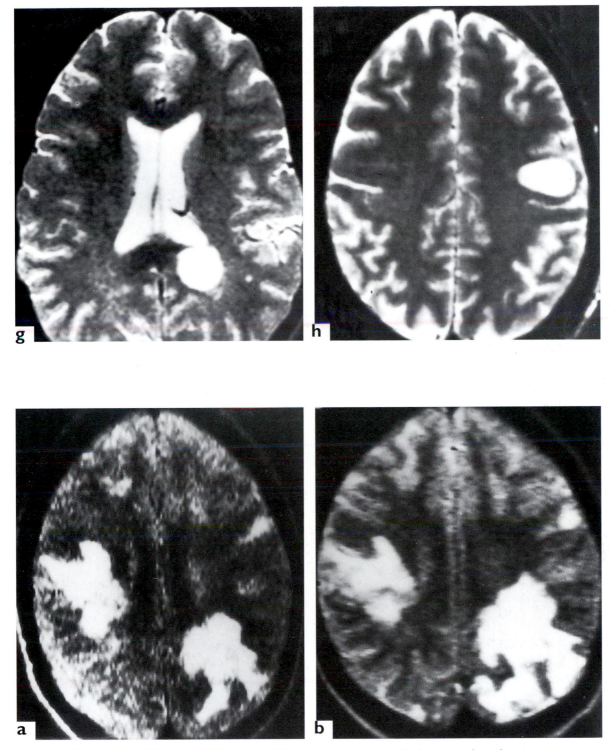

Figure 101 Axial MRIs of ADEM in a 17-year-old girl who had a grand mal seizure accompanied by aphasia. Recovery was complete within 24 hours. MRI (a) showed enormous bilateral areas of increased signal intensity (AISI). The patient was asymptomatic for a year, until she had another seizure after discontinuing her anticonvulsant medication. Her neurological examination was normal, but MRI showed the same AISI. Three years later, the follow-up MRI (b) in this completely asymptomatic subject still showed the same AISI

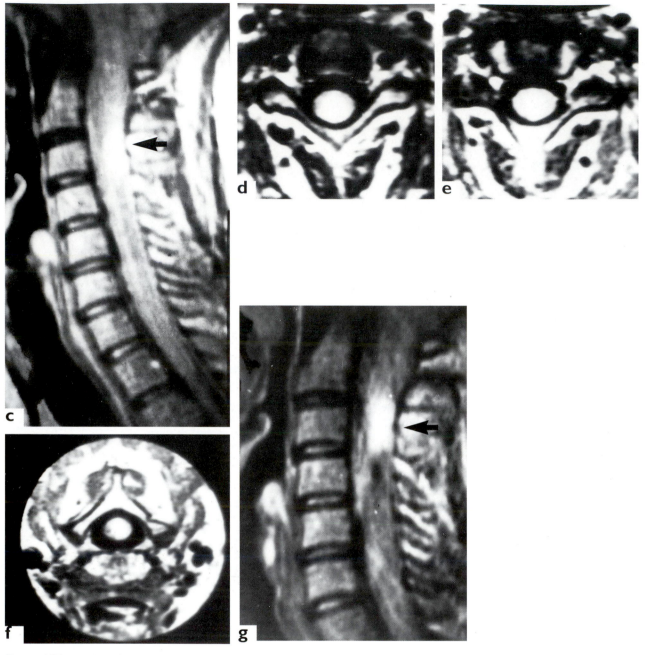

Figure 101 continued

A 42-year-old woman started complaining of rapidly progressive clumsiness of her hands. The condition persisted unchanged until, after 2 months, she was examined neurologically and found to have complete loss of position and vibration senses with severe pseudoathetosis of the arms and hands. MRIs of the spinal cord (c–g) revealed an area of increased signal intensity (AISI) at the level of C3 (arrows). Treatment with corticosteroids resulted in complete clinical recovery. On MRI a year later (g), the C3 AISI (arrow) remained unchanged despite the complete absence of neurological symptoms. It is difficult to understand how these AISI can be called 'lesions' when they cause no neurological signs or symptoms. Their true nature remains in dispute. Such 'lesions' call into question the concept of 'burden of disease' commonly used to describe the extent of AISI in patients with multiple sclerosis. From Poser, 1989; reproduced with the permission of Elsevier Science BV

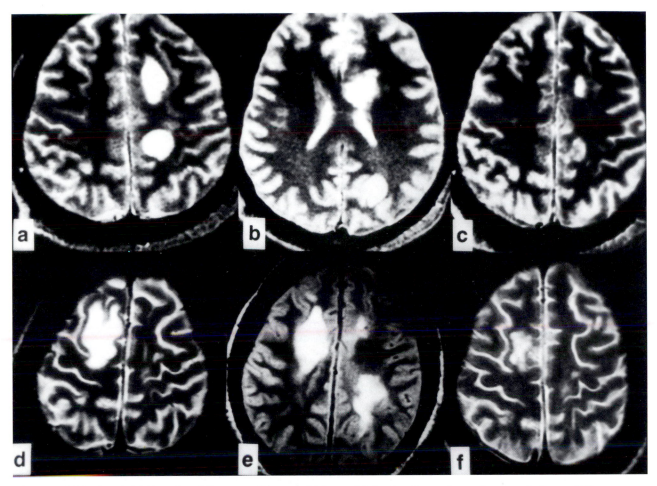

Figure 102 Axial MRIs of a 29-year-old man with multiphasic disseminated encephalomyelitis (MDEM). The patient had a rapidly progressing right hemiparesis. The initial scan (a) revealed two large AISI in the left hemisphere, and a small one in the right hemisphere posteriorly. Biopsy of the left frontal AISI revealed inflammation and demyelination. AISI of this size are typical of ADEM. The patient recovered without treatment within 4 months. A follow-up scan taken 6 months later (b) showed shrinkage of the AISI and, 1 year later (c), two small AISI were all that remained.

Three years later, following a flu-like illness, the patient complained of a moderately severe left hemiparesis. MRI at that time (d) showed a large new AISI in the right frontal area. A follow-up study 2 weeks later (e) showed what appeared to be reactivation of the two left hemisphere AISI despite the absence of right-sided symptoms. The patient recovered after treatment with methylprednisolone. A final scan 3 months later (f) showed considerable regression of all AISI.

This case is a good example of the value of MRI in ruling out multiple sclerosis in a patient whose history is compatible with such a diagnosis. Courtesy of Dr Saeed Bohlega, King Faisal Specialists Hospital, Riyadh, Saudi Arabia; modified from Khan *et al.*, 1995

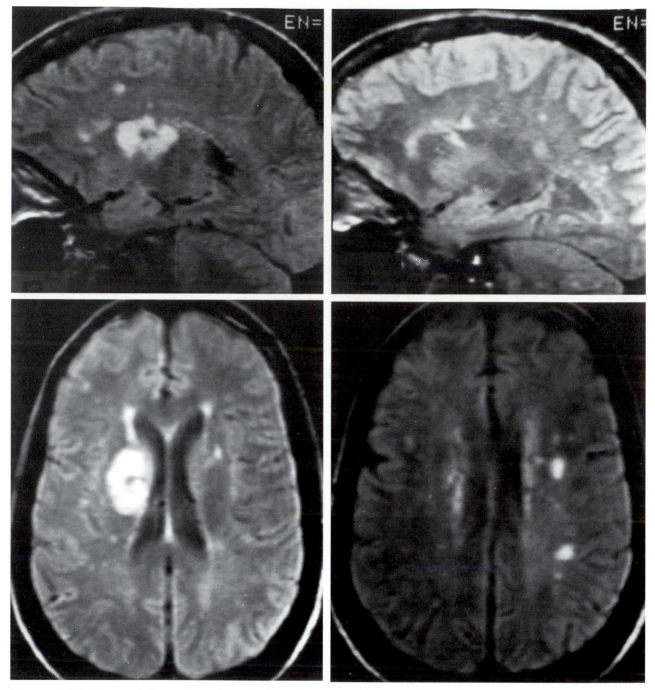

Figure 103 Coronal (upper) and axial (lower) proton-density MRIs of a patient with Lyme disease show large periventricular areas of increased signal intensity similar to those seen in ADEM (see also Figure 94). Courtesy of Dr Patricia Hibberd, Massachusetts General Hospital, Boston, MA

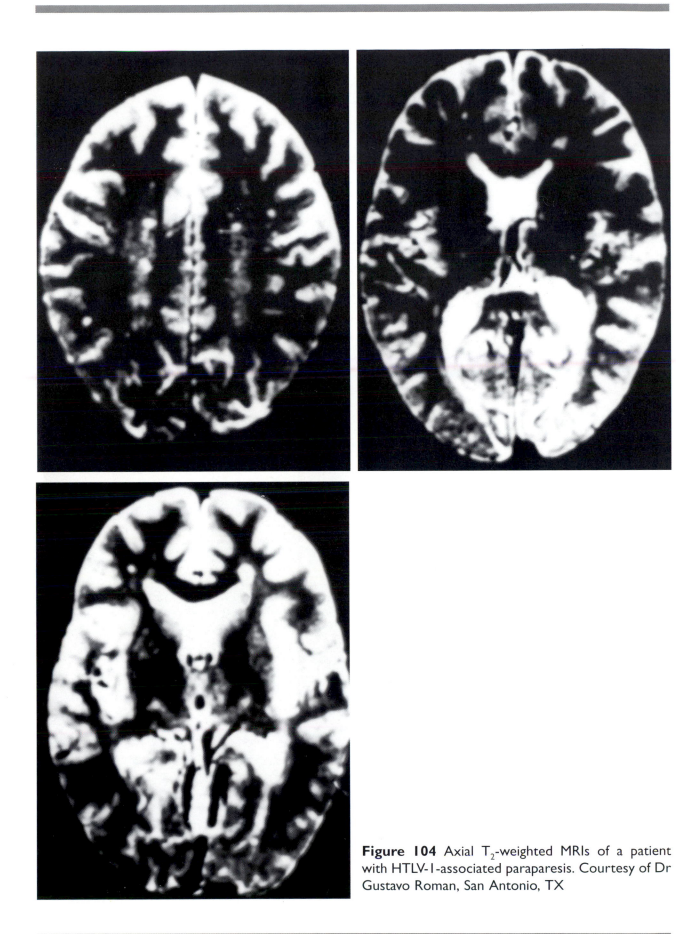

Figure 104 Axial T_2-weighted MRIs of a patient with HTLV-1-associated paraparesis. Courtesy of Dr Gustavo Roman, San Antonio, TX

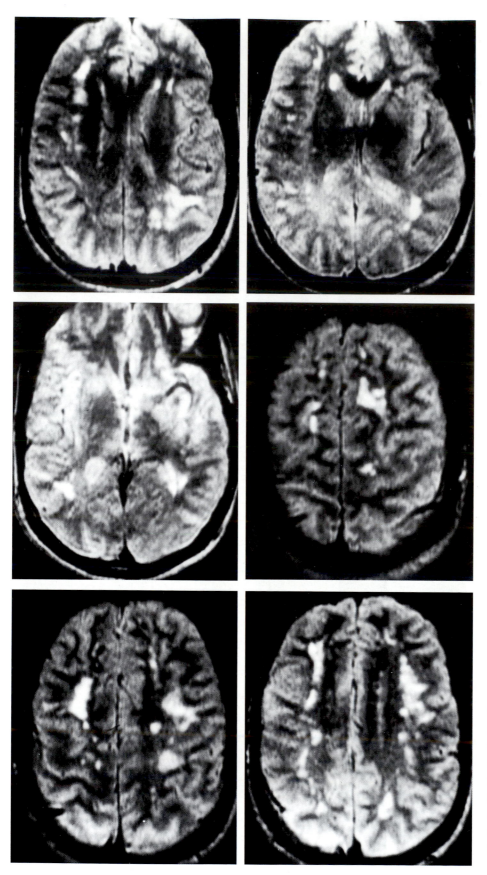

Figure 105 Axial proton-density MRIs of a patient with cerebral AIDS. Nearly all of the lesions lie at the periphery rather than in the periventricular area; this pattern is similar to that seen in ADEM (see Figure 100a)

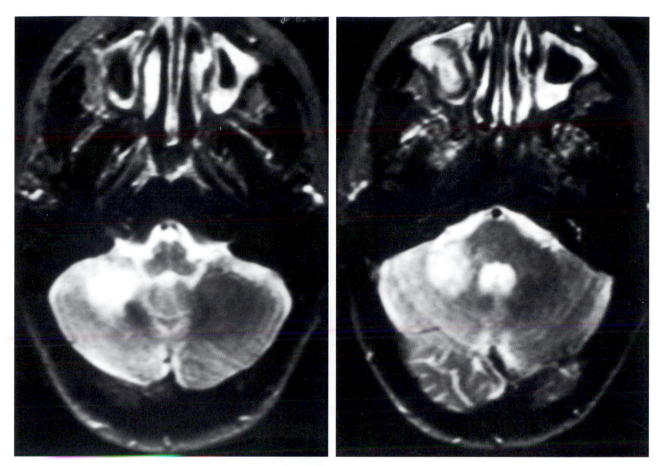

Figure 105 continued
Axial T$_2$-weighted MRIs showing a single cerebellar lesion (see also Figure 96)

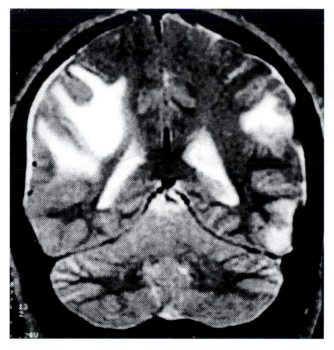

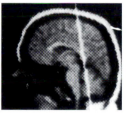

Figure 106 Coronal T$_2$-weighted MRI of a patient with neurosarcoidosis. The lobar areas of increased signal intensity are similar to those seen in ADEM (see also Figure 99)

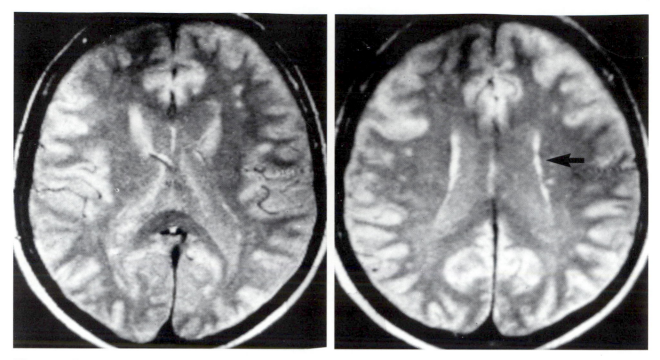

Figure 107 Axial proton-density MRIs of a patient with hypertensive cerebrovascular disease (Binswanger's) show scattered punctate and linear periventricular areas of increased signal intensity (arrow)

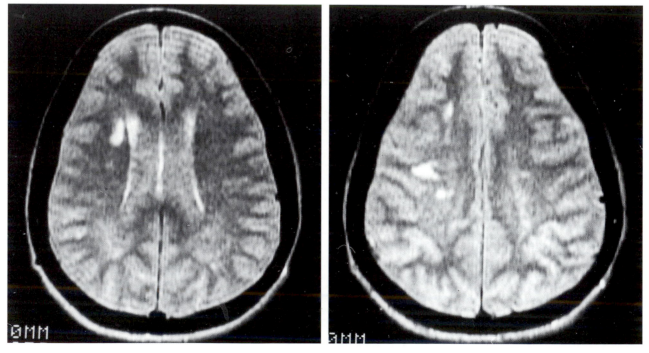

Figure 108 Axial proton-density MRIs of a patient with inflammatory vasculitis. The pattern of the areas of increased signal intensity as well as the clinical history are similar to those commonly seen in multiple sclerosis. Courtesy of Dr Michael Ronthal, Beth Israel Deaconess Medical Center, Boston, MA

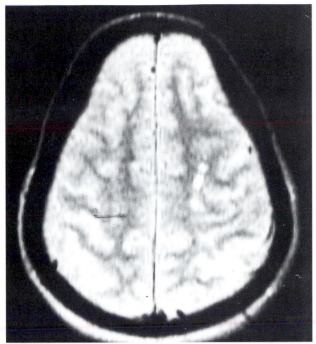

Figure 108 continued
Axial proton-density MRI of a patient with inflammatory vasculitis

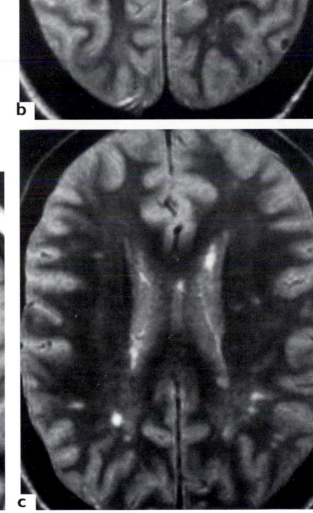

Figure 109 T$_1$-weighted MRI using gadolinium–EDTA injection (a) and proton-density MRIs (b & c) of a patient with chronic fatigue syndrome (CFS). An enhancing lesion in the right parasagittal frontal cortex (a, arrow) corresponded to the patient's left-sided Babinski's sign. Despite the presence of neurological abnormalities, the patient fulfilled the diagnostic criteria for CFS

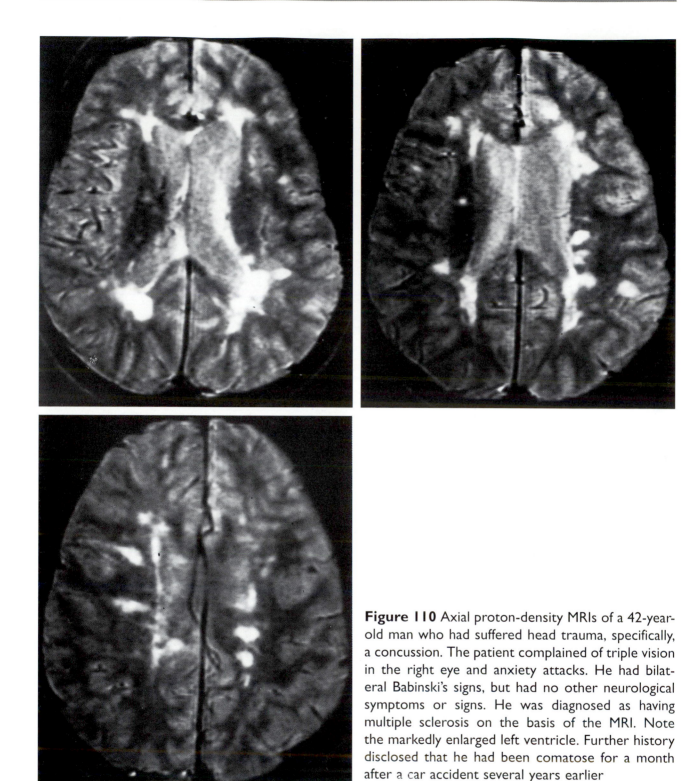

Figure 110 Axial proton-density MRIs of a 42-year-old man who had suffered head trauma, specifically, a concussion. The patient complained of triple vision in the right eye and anxiety attacks. He had bilateral Babinski's signs, but had no other neurological symptoms or signs. He was diagnosed as having multiple sclerosis on the basis of the MRI. Note the markedly enlarged left ventricle. Further history disclosed that he had been comatose for a month after a car accident several years earlier

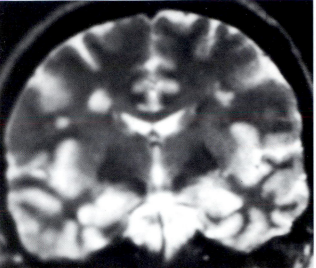

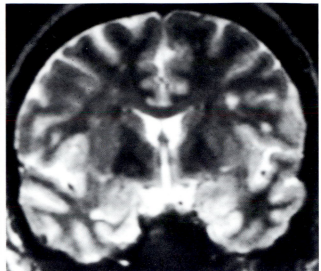

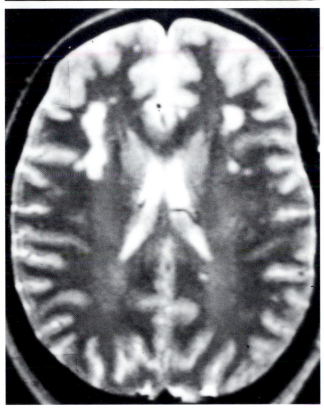

Figure 111 Coronal (upper) and axial (lower) T$_2$-weighted MRIs of a patient who had suffered a mild concussion without loss of consciousness. His only complaint was headache. Multiple areas of increased signal intensity are present in the cerebral white matter

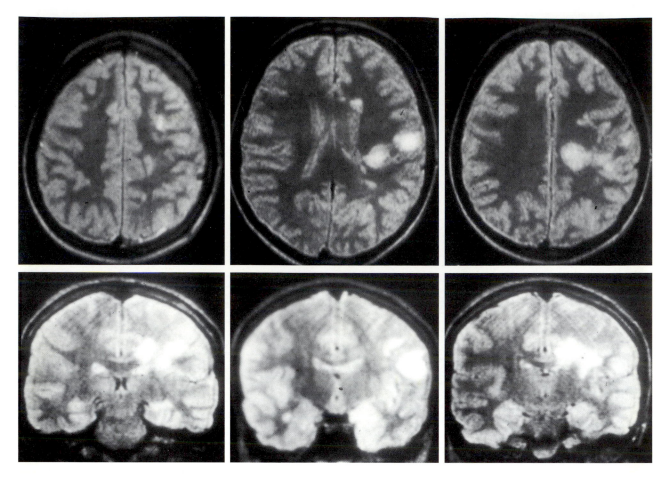

Figure 112 Axial (upper) and coronal (lower) proton-density MRIs of a patient who had complicated migraine. The pattern is similar to that seen in ADEM. Large cortical areas of increased signal intensity can be seen

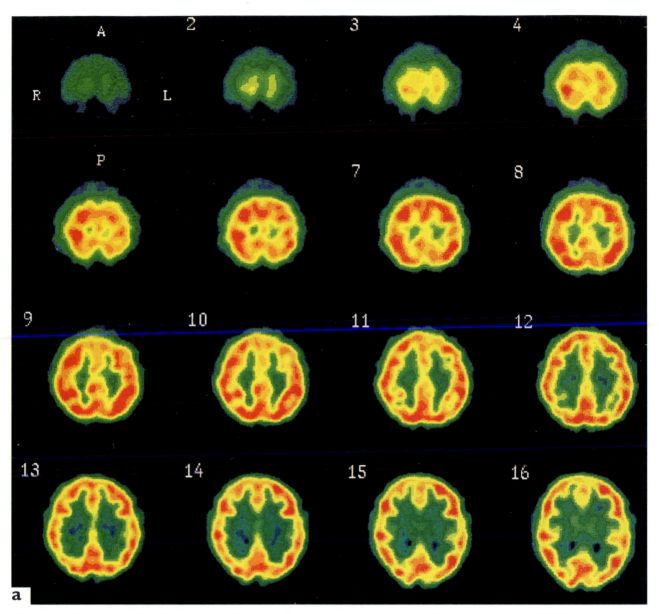

Figure 113 Axial SPECT (continued overleaf) of the brain, using ethylcystinate-dimer (ECD), of a 60-year-old man with chronic progressive multiple sclerosis in a recent-onset acute confusional state. There is bifrontal hypoperfusion extending into the parietal lobes. Courtesy of Dr Barbara Pickut, Middelheim General Hospital, Wilrijk–Antwerpen, Belgium

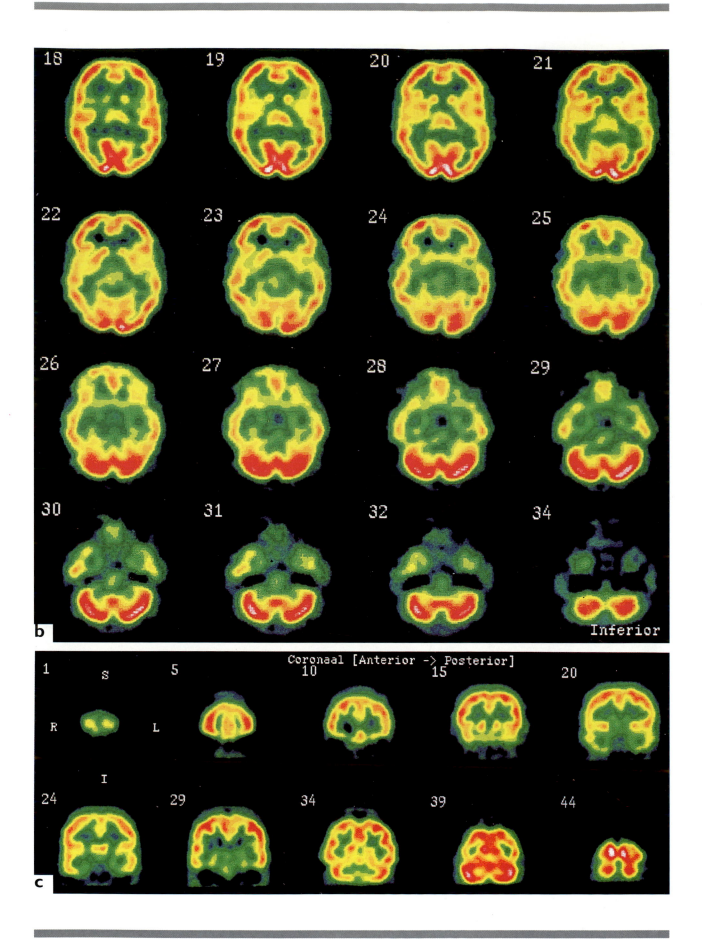

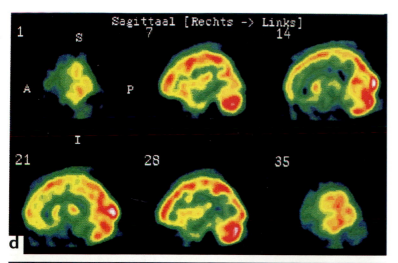

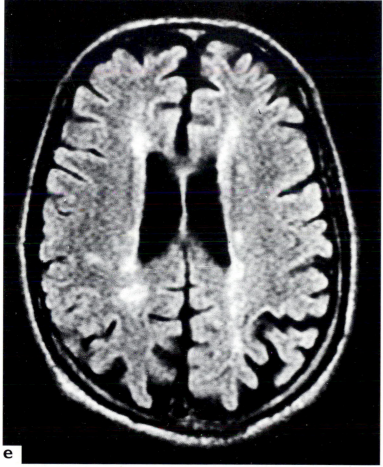

Figure 113 continued
Axial (b), coronal (c) and sagittal (d) ECD SPECT, and axial proton-density MRI (e), of the brain of a 60-year-old man with chronic progressive multiple sclerosis in a recent-onset acute confusional state. Courtesy of Dr Barbara Pickut, Middelheim General Hospital, Wilrijk–Antwerpen, Belgium

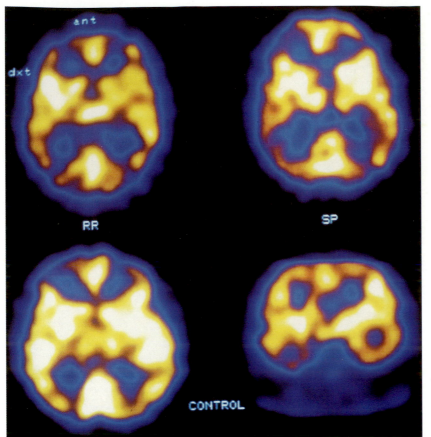

Figure 114 SPECT of the brain, using technetium-99 hexamethyl-propyleneamine oxime, of two multiple sclerosis patients (upper) compared with a control (lower). Both patients clearly show a reduction of perfusion. Courtesy of Dr Jan Lycke, Sahlgrenska University Hospital, Göteborg, Sweden. For technical details, see Lycke et al., 1993

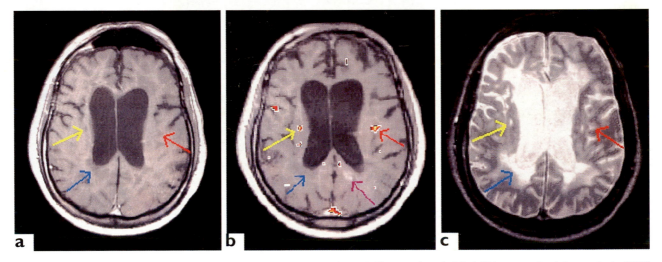

Figure 115 Axial post-gadolinium T_1-weighted (a & b) and T_2-weighted (c) MRIs, overlaid by cobalt–PET displays. Areas of non-enhancing hypodensity in seen in a (arrow). Cobalt–PET-enhanced lesions may coincide (yellow and red arrows), but long-standing confluent lesions are neither enhanced nor show cobalt uptake (blue arrows). There is an enhanced lesion in b (magenta arrow) which does not show cobalt uptake. In general, there is good, but not complete, correlation between disease activity (gadolinium-enhanced) and cobalt–PET-enhanced lesions. Unpublished results, courtesy of H. Jansen, D. de Coo, J. De Reuck, J. Korf and J. Minderhoud, Groningen, The Netherlands, and Ghent, Belgium; courtesy of Dr Jakob Korf, University Hospital, Groningen, The Netherlands. For technical details, see Jansen et al., 1995

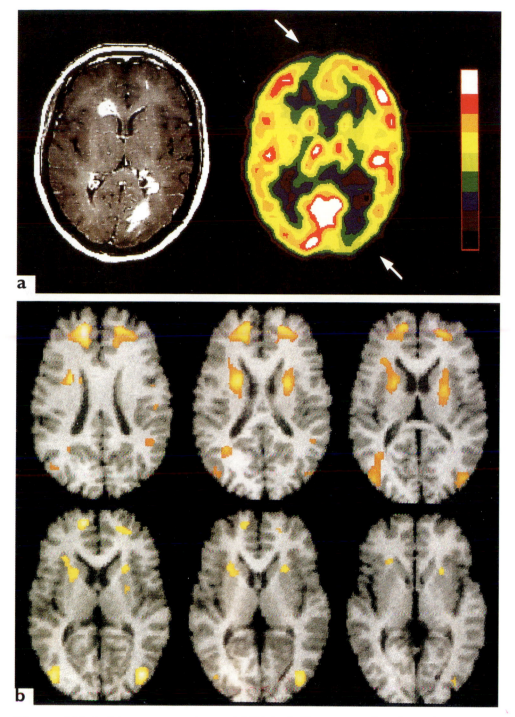

Figure 116 Axial contrast-enhanced MRI (a, left) compared with an ^{18}F-labeled deoxyglucose (FDG)–PET scan (a, right) of the brain. The color scale represents glucose consumption from 0 (black) to 45 (white) mol/100 ml/min. Areas of abnormal cortical glucose consumption (arrows) may be considered a consequence of the subcortical demyelinating lesions seen on the MRI. The significantly ($p = 0.005$) lower differences of glucose metabolism in multiple sclerosis patients with ($n = 19$) vs without ($n = 16$) fatigue are superimposed on axial MRIs of a healthy subject (b). Differences are most prominent in the frontal cortex, striatum and adjacent white matter. Courtesy of Drs U. Roelcke and K. Leenders, Paul Scherrer Institute, Villigen, Switzerland; from Roelcke et al., 1997; reproduced with the permission of Lippincott–Raven (b)

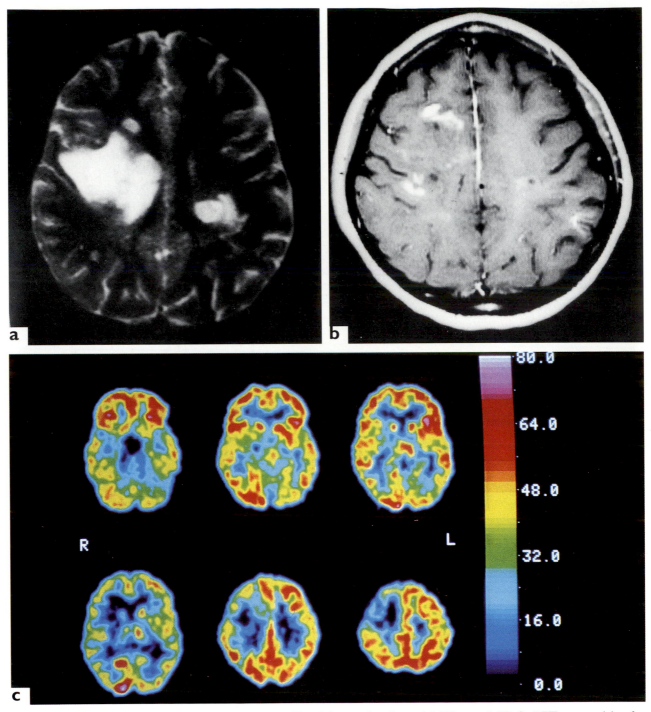

Figure 117 T$_2$- (a) and gadolinium-enhanced T$_1$- (b) weighted axial MRIs, and FDG–PET scans (c) of a patient with Baló's disease show marked depression of glucose utilization in the white matter and overlying cortex of the right frontal lobe corresponding to the areas of increased signal intensity seen on MRI. In the affected right cortex, glucose metabolism was approximately 30% lower than on the left, and was also reduced in the right striatum, thalamus and parahippocampal region. As a remote effect, glucose utilization was diminished in the left hemisphere of the cerebellum. Slight glucose hypometabolism is also apparent in the left visual and centroparietal cortex. Courtesy of Dr Rudiger Seitz, Heinrich-Heine University, Dusseldorf, Germany. For technical details, see Hanemann *et al.*, 1993

Index